Politics of Eugenics

This book analyzes whether the "new debate on genetics" owes a debt to eugenic practices by welfare democracies of the 1930s and 1940s. More specifically, the question is whether precisely the same "eugenic rationale" used in the 1930s is philosophical akin to a new rationality unfolding in some Western European welfare societies that find themselves trapped in the modern dilemma of choosing between increasing immigration and population growth that leads to economic prosperity on the one hand, or halting immigration, protecting national identity, and suffering economic stagnation on the other.

By analyzing policies of integration and assisted reproduction technology (ART) in Northern European nation states such as Sweden, Finland, Denmark as well as in Israel, we find a historical continuity between "old eugenics" and current reproductive and family planning subsidies and integration policies. By focusing on the concept of welfare productionism, we trace a continuing rationale between the eugenic policies of the past and current investments of ART. These programs are rationalized as universal programs for the whole of the population. However, in this book the authors suggest that they served the goal of reproducing a productivist, national middle class which are enticed to reproduce.

This work will be of great interest to students and scholars of racism, extremism, European politics, population politics, and the social impact of science and technology.

Alberto Spektorowski is a Senior Lecturer in Political Science at Tel Aviv University, Israel.

Liza Ireni-Saban is an Assistant Professor in the Lauder School of Government, Diplomacy and Strategy at the Interdisciplinary Center Herzliya, Israel.

Routledge studies in extremism and democracy

Series Editors: Roger Eatwell

University of Bath

and

Matthew Goodwin

University of Nottingham.

Founding Series Editors: Roger Eatwell

University of Bath

and

Cas Mudde

University of Antwerp-UFSIA.

This new series encompasses academic studies within the broad fields of "extremism" and "democracy." These topics have traditionally been considered largely in isolation by academics. A key focus of the series, therefore, is the (inter-)*relation* between extremism and democracy. Works will seek to answer questions such as to what extent "extremist" groups pose a major threat to democratic parties, or how democracy can respond to extremism without undermining its own democratic credentials.

The books encompass two strands:

Routledge Studies in Extremism and Democracy includes books with an introductory and broad focus which are aimed at students and teachers. These books will be available in hardback and paperback. Titles include:

Understanding Terrorism in America
From the Klan to al Qaeda
Christopher Hewitt

Fascism and the Extreme Right
Roger Eatwell

Racist Extremism in Central and Eastern Europe
Edited by Cas Mudde

Political Parties and Terrorist Groups (2nd Edition)
Leonard Weinberg, Ami Pedahzur and Arie Perliger

The New Extremism in 21st Century Britain
Edited by Roger Eatwell and Matthew Goodwin

New British Fascism
Rise of the British National Party
Matthew Goodwin

The End of Terrorism?
Leonard Weinberg

Mapping the Extreme Right in Contemporary Europe
From local to transnational
Edited by Andrea Mammone, Emmanuel Godin and Brian Jenkins

Varieties of Right-Wing Extremism in Europe
Edited by Andrea Mammone, Emmanuel Godin and Brian Jenkins

Routledge Research in Extremism and Democracy offers a forum for innovative new research intended for a more specialist readership. These books will be in hardback only. Titles include:

1 **Uncivil Society?**
Contentious politics in post-Communist Europe
Edited by Petr Kopecky and Cas Mudde

2 **Political Parties and Terrorist Groups**
Leonard Weinberg and Ami Pedahzur

3 **Western Democracies and the New Extreme Right Challenge**
Edited by Roger Eatwell and Cas Mudde

4 **Confronting Right Wing Extremism and Terrorism in the USA**
George Michael

5 **Anti-Political Establishment Parties**
A comparative analysis
Amir Abedi

6 **American Extremism**
History, politics and the militia
D. J. Mulloy

7 **The Scope of Tolerance**
Studies on the costs of free expression and freedom of the press
Raphael Cohen-Almagor

8 **Extreme Right Activists in Europe**
Through the magnifying glass
Bert Klandermans and Nonna Mayer

9 Ecological Politics and Democratic Theory
Mathew Humphrey

10 Reinventing the Italian Right
Territorial politics, populism and 'post-Fascism'
Carlo Ruzza and Stefano Fella

11 Political Extremes
An investigation into the history of terms and concepts from antiquity to the present
Uwe Backes

12 The Populist Radical Right in Poland
The patriots
Rafal Pankowski

13 Social and Political Thought of Julius Evola
Paul Furlong

14 Radical Left Parties in Europe
Luke March

15 Counterterrorism in Turkey
Policy choices and policy effects toward the Kurdistan workers' party (PKK)
Mustafa Coşar Ünal

16 Class Politics and the Radical Right
Edited by Jens Rydgren

17 Rethinking the French New Right
Alternatives to modernity
Tamir Bar-On

18 Ending Terrorism in Italy
Anna Bull and Philip Cooke

19 Politics of Eugenics
Productionism, population, and national welfare
Alberto Spektorowski and Liza Ireni-Saban

Politics of Eugenics

Productionism, population, and national welfare

Alberto Spektorowski and Liza Ireni-Saban

Routledge
Taylor & Francis Group
LONDON AND NEW YORK

First published 2014
by Routledge
2 Park Square, Milton Park, Abingdon, Oxon OX14 4RN

Simultaneously published in the USA and Canada
by Routledge
711 Third Avenue, New York, NY 10017

Routledge is an imprint of the Taylor & Francis Group, an informa business

British Library Cataloguing in Publication Data
A catalogue record for this book is available from the British Library

Library of Congress Cataloging in Publication Data
Spektorowski, Alberto,
Politics of eugenics: productionism, population, and national welfare / Alberto Spektorowski and Lisa Saban.
pages cm. – (Routledge studies in extremism and democracy; 19)
Includes bibliographical references and index.
1. Eugenics. 2. Biopolitics. 3. Emigration and immigration–Government policy. 4. Social policy. 5. Multiculturalism. I. Saban, Lisa. II. Title.
HQ751.S677 2013
363.9'2–dc23
2013002753

ISBN: 978-0-415-81431-7 (hbk)
ISBN: 978-0-203-74023-1 (ebk)

Typeset in Times New Roman
by Wearset Ltd, Boldon, Tyne and Wear

Printed and bound in the United States of America by Publishers Graphics, LLC on sustainably sourced paper.

Contents

Introduction

In this book we analyze an item that most people in liberal societies would wish to obscure, and even deny. There is no question that a backlash against liberal multiculturalism is sensed in most Western liberal democratic societies. It is clear to all that an increasing number of voices in these societies are being raised against immigration and in favor of what could roughly be defined as the defense of national culture and the national stock. With this background, several liberals might ask themselves whether it is possible that modern democratic societies are beginning to review social engineering policies reminiscent of eugenic policies of the past.

On September 23, 2008, John LaBruzzo, a Republican from Metairie, Louisiana, raised a heated debate by offering \$1000 to women receiving government welfare if they willingly chose to be sterilized. According to LaBruzzo, "People receiving government aid such as food stamps and publicly subsidized housing are reproducing at a faster rate than more affluent, better educated people who presumably pay more tax revenue to the government."[1] Although such an insensitive statement has no place in liberal democracy, it reveals that eugenics was never actually removed from the world.

In a conference on the world financial crisis at the School of International and Public Affairs, Columbia University (February 2009), Professor Joseph Stiglitz notes in relation to Western democracies' management of the financial system: "...Western capitalist societies affected by the world financial crises have started to nationalize banks. However, it is important for liberal governments not to define nationalization as 'nationalization'." Other concepts are used to define a phenomenon whose effects in the psychology of economic activity might be negative.

In the same vein, there are concepts in the realm of health and welfare policies that liberals in open, multicultural societies would hardly want to hear, such as population problems, national hygiene, "national productionism," and eugenics. Most analysts worldwide would not have much difficulty relating the ideas of eugenic policies and national productionist ideas to the Nazi and Fascist regimes.

Indeed, as a mixture of science and policy, as a discipline and social movement, eugenics lies at the interface of biological science and society.[2] This so-called science, therefore, seeks to regulate human procreation by encouraging the fecundity of allegedly genetically superior groups in society, while discouraging "defectives" from producing children, since they would replicate their deficiencies. Thus, it is easy to understand why totalitarian or racist movements were enticed by eugenic science. However, liberals would find it extremely difficult to accept these concepts and methods as a part of democratic practices in current times. More difficult yet is to accept that some of these policies, disguised and covered under genetic legitimacy, are reappearing in current public discourse during times of economic competition and pressure of immigration. Let us highlight the dilemma.

Most current liberal democracies are multicultural and market oriented. Advocates of immigration processes use utilitarian arguments to argue that since Western democracies' rate of fertility is low, immigration appears as one of the engines of economic development. Immigrants' contribution to the impoverished arks of Western pension funds can explain why the latter still stand afloat. At the same time, however, differently from the United States, immigration constitutes a problem for several Western democracies especially in Europe, which still considers massive immigration as a challenge to their cultural identity makeup (Gradstein and Schiff 2006). To be precise, the United States in the past has run into intense eugenic policies. As Hansen and King explain, despite the powerful myth of the United States as a country of immigration, the United States adopted eugenic-based immigration policies in the 1930s (Hansen and King 2001: 247; Lombardo 2008). The dominant political elites at that time supported restrictive immigration policies based on the scientific hierarchy of races.[3]

There was widespread perception that certain races entering the United States were inferior, and would threaten the American gene pool. However, that perception has substantially changed, and although debates focus on what to do about illegal immigration, the basic idea of liberals is that legal immigration, regardless of race, is economically productive. Indeed, immigration constitutes an integral part of the ethos of liberal democratic societies that display a special vocation for free market economics, such as the United States. That makes American society a liberal productionist welfare society. Immigration, however, is hardly the ethos of most European societies, which consider themselves nation states, or that at least pay special attention to the question of national identity. These nations strive to preserve their welfare society and their national culture. The ongoing debate over the tendency among minority nationalisms in binational or multinational countries of the European Union to split, boosts the idea that small culturally homogeneous countries are also economically viable, because homogeneity and productivism could be considered tied concepts. In this sense, it is not out of the question to suppose that the Scandinavian model of liberal welfare ethno-nationalism may be considered a good example to follow. Furthermore, despite the assumption that capitalist globalization has undermined both social democracy and the nation state, there is an increasing perception that globalization boosts the necessity of a welfarist productionist system. There is the conviction in Sweden and other Scandinavian countries that social-democratic welfare systems work best, to the extent they do work, in ethnically and culturally homogeneous (and preferably small) nations whose citizens, treating each other as members of an extended family, are loath to exploit government provisions for the needy. It is thus quite clear that some of the latter nations, which have become multicultural de facto, still feel reluctant to accept themselves as such.

For these types of European societies two factors—immigration and the economy—are intimately connected. However, due to the fact that they are both national and welfarist, they are trapped with the need for more working hands, which only immigration could provide. However, due to the universal and benevolent welfare system, a huge share of immigrants hardly work, becoming for that reason a burden for the European welfare state. In this last sense, what really happens is that a crowd of immigrants are kept in third world pockets, supported by a welfare state that barely gets a contribution from them. From the perspective of both national identity and economic productivism, the system does not work. In this sense, if they want to take advantage of the productivist side of immigration they have to drop their welfare system. However, at the same time, if political leaders fear that an increasing population of immigrants might jeopardize national identity, which is intimately connected to the welfare system, they might choose to stop immigration. In that case, they will suffer a drastic reduction of working hands and low levels of production, or at least they will not benefit from the productivist contribution of immigration.[4] To this point it is clear that if European societies consider that they have to protect their national culture and run a universal welfare system, it is important to note that this will hardly work in a global dynamic economy and will not work with immigrants.

Two strategies, which is some cases are complementary and in some are contrasting, can address this new situation. One is civic integration and the second a redefinition for modern times of positive eugenics.

European political elites in the last decades pursued an American strategy featuring a neoliberal concept of civic integration. That determination coincided with the opening of the markets and increasing mobilization of population within European frontiers. Indeed, with the enlargement of the European Union and the Schengen agreements, which provide for open frontiers, Europeans could move from region to region, guaranteeing the constant mobilization of working hands where more are needed.[5]

However, the central question was whether the newcomers would integrate into society. Indeed, as noted, European countries are becoming sensitive not only to the necessity of the market but especially to their cultural identity, leading to increased interest in the language of civic integration (Joppke 2004). Civic integration and neoliberal economics were perceived as corresponding concepts. As Christian Joppke suggests, the logic of civic integration is to treat migrants as individuals who are depicted as responsible for their own integration. In that sense civic integration is an extension of the austere neoliberalism that frames economic globalization (Joppke 2007). A migrant's initial entry into the new society is thus precarious and different from the past, and the migrant must gradually "earn" the rights of full (and this means, above all, permanent) membership.[6] Following this line of reasoning integration could be considered as an instance, next to eugenics in the past and workfare today, of "illiberal social policy" in a liberal state. According to Desmond King, these productionist ideas are embedded in liberal ideology itself. Thus, liberalism's core tenets of freedom and equality presuppose that "members of the polity possess the necessary reasoning powers or ability to ... plan for their future" (King 1999: 8 fn 19). There is an ongoing debate in several Western European democracies, such as France, Great Britain, Holland, and Denmark, in which moral and utilitarian arguments are used to advocate civic integration. Some analysts conclude that the programs of civic integration are extremely costly and do not ensure returns (Pierson 2001; Taylor-Gooby 2004). In other words, it seems that the demands to shift from liberal contractualism to duties embedded in civic integration of immigrants are still extremely costly. Logically, thus, if civic integration is a next step to eugenics as assumed by King, and hardly works, what comes next is the adaptation of eugenics for modern liberal times.

Indeed, we argue that underneath the civic integration debate other venues to deal with the necessity of population growth are being examined; one of these is related to the nexus eugenics–genetics.

There is no question that the recent increase in availability of genetic technologies may provide a surreptitious way for governments to control the demographic makeup of their countries. Duster (2003) for example, argues that reproductive decision-making is not a matter of individual choices but may be rather a state-enforced policy. That is the reason why some critics consider government subsidization of genetic technologies for determined groups of the population to be the necessary resurrection for eugenic population policies of the first half of the twentieth century. Others, however, strongly claim that there is no way to compare the "old" eugenics and current genetic subsidization (Buchanan *et al.* 2000; Bunton and Petersen 2005; Lemke 2002; Mannion 2006).

Thus, the analysis presented in this book joins other studies that characterize the gradual transformation in bioethics by an increasing appeal to moral and ethical values and norms that should guide the use of new genetic technologies: What are the ethical groundings of the new genetic technologies' normative discourse? Do the current policies of genetic technologies live up to their moral claims and standards?[7] While the existing bioethics literature evaluates the normative power of new genetic technologies, this book treats the normative discourse as a device to legitimize the state control over diverse actors, both internally (citizens) and externally (states). The nature of this interference differs depending on the governance system of genetic technologies. Precisely because of that, we pay attention to Michel Foucault's concept of biopower. Indeed, the way discourses in bioethics are embedded in politics should be framed with the Foucauldian terminology of biopower. In short, in biopower, Foucault referred to certain power/knowledge techniques to rationalize and manage the "life" of a population. This technique emerged in the second half of the eighteenth century, when population gained prominence in various discourses of how to police and control extended groups of human beings in order to empower larger institutional agents (such as the nation state). The new regime of biopower has generated arguments and claims of bioethics that respond to population policy problems.

The wide discourse of bioethics on the possible continuity of eugenics in contemporary human genetics could be explained by biopower as inspired by different kinds of "bios." The term bio within the biopolitical underpinning of the new genetics is not to be couched in explicitly biological terms, but rather an understanding of the population at risk. Population discourse in such terms is seen as something that is threatened by internal and external agents, and which can survive and be improved by the elimination of those threats. The kinds of bios that have been discussed in the present book in relation to eugenic-based population policy are racial and national. It is argued that the new bioethics needs to engage in the political discourse of technologically developed people's-home countries in which productivist social democratic and labor movements pursue eugenic exclusionary practices in order to preserve the hegemonic national community under levels of production that allow their economic survival.

Thus this book follows the "continuity hypothesis"—and suggests that today's genetic subsidies are embedded in the modernist rationale of early twentieth century eugenics. Its most important feature is to encourage the reproduction of the productive sectors of the population that are of the dominant ethnic group. In this sense it could be defined as an "affirmative action of the strong," in which the elite is encouraged to reproduce itself in order to mobilize the whole of society forward, while at the same time the social margins, generally alien ethnic groups, are encouraged to not reproduce. In short, this means welfare for the strong, which implies investment in manpower for economic progress and not necessarily investment for the needy.

Scholars who support the claim that eugenics is still with us accept Daniel Kevles' claim that eugenics has shifted from "mainline" to "reform" eugenics. Other scholars, such as David Gems (1999) and Alexandra Plows (2011), discern the libertarian reform character of current reformist eugenics. They claim that current genetic practices such as genetic testing and screening are characterized by free choice. Accordingly, current free "choice" is the key that allows differentiation of scientific individualistic reform eugenics from the unscientific collectivist mainline eugenics of the past.

While this book agrees that current scientific individualistic eugenics differs from the mainline eugenics of the past, still it stresses the connecting line between "mainline" and "reformist" eugenics in the past, as well as the more individualistic reformist eugenics of current times. Indeed, "reform eugenics" is not free from some of the diseases of "mainline eugenics." Social Darwinian ideas penetrated social reformism, especially productivist social reform, which found expression in the policies of social engineering and in strategies designed to enhance strong populations. Despite current individualist features, we suggest that the ideas of reformist eugenics embedded in the ideology of welfare socialist productionism fit well and are enticing to current welfare nation states. For this reason, the attempt to search for the continuing line between the background of eugenic welfare productionism and current assisted reproduction technology (ART) policies is the right path to follow.

As we shall see, the ideology of welfare-productivist socialism, as expressed by the British Fabians, the Swedish and Danish Social Democrats, Finnish nationalists and Social Democrats, and Israel Laborites, is directly influenced by social Darwinism. As such they set a middle road between the total exclusionism of racism on the one hand and liberal inclusionism on the other, and between authoritarian fascism and democratic nationalism.

However, the important point is that the welfare-productivist socialism we deal with is akin to social democracies in nation states or in countries fostering to protect their national identity and their economic stability. In other words, we stress that while eugenics was generally used across the board by liberals, socialists, and racists, it was specially tailored for social democratic or progressive productivist parties in nation states. Several socialist movements, and ideologues such as the British Fabians, adopted a very clear and simple formula. If a socialist society wants to survive as a welfare society, its productive elements should be enhanced while reducing the feeble margins of society. Social democrats and laborites in the 1930s thus endorsed eugenics covered by the concept of "productive welfarism" with the hope of creating a state consisting only of those who positively contributed to the nation. While at first sight they might be compared with Nazi racist totalitarianism, in reality, social democrats did not believe in racial superiority, though they did believe that the only way a welfare state can survive is by reproducing the high quality stock (Schmidt 2000, 2002). While Nazis preferred racist purity to productionism, social democrats synthesized universal and nationalist values, and were eager to preserve national democratic culture. In other words, rather than racist, these movements defended a democratic type of nationalism while promoting state intervention in social and economic affairs. In more senses than one, they promoted an exclusionist version of democratic inclusion. The meaning of this is that inclusion is conditioned to the hegemony of a productionist social culture based as noted in welfare for the strong and in reduction of the social margins.

In short, despite observers who may argue that distinctions between Nazi racism, fascism, and productivist socialism are artificial and difficult to draw, we believe that it is important to distinguish between them. However, we claim that by remarking on the shift from racial to "national-productivist hygiene," this book refurbishes a political category that synthesizes national welfare, productivism, and eugenics, and that we redefined as "welfare productionism." A eugenics-based welfare productionism thus constitutes an intermediate category between racism and liberalism. Productivist socialism thus fell short of Nazi racism and of fascist authoritarianism, and it is precisely this non-racist welfare productionist ideology that sets the connecting links between the eugenics of the past and genetics in present times. Productivist socialists endorse "reformist eugenics" in order to engineer a welfare society for all, blacks and whites, Jews, Christians, nationals, and foreigners. In this sense they are hardly racist; nonetheless the concept of productionism implies a conditioned acceptance of "the other" that could be defined as an exclusionist type of inclusionism.

Inclusion is designed for the stronger and modernized sectors of society. Paradoxically, groups that appeared feeble in the past such as women, homosexuals, and lesbians could be portrayed as the new "strong" sectors of modern society.

Thus, it is not strange that modern societies would recommend genetic treatment for lesbians and homosexuals, who generally belong to the modern national ethnic group. Paradoxically, although discriminated against for years and discriminated against nowadays, these groups are becoming essential for the survival and quality of the nation. While for right-wing racists in the past, and religious and conservatives in present times, homosexuals are portrayed as degenerate, for progressive welfare productivists they are an important engine of the productive country that strives to preserve its societal culture.

It is not strange that economic aid will be provided to single mothers, while at the same time the gates to immigration are increasingly being closed and "immigrant" or non-national ethnic minorities are discouraged from breeding.

To put it in more simplistic terms, liberals consider it morally compelling to support the "right of homosexuals and lesbians" to raise families and materialize their "right to breed." Governments, as we shall see, have endorsed these "progressive policies." At the same time, whether consciously or not, these policies fit well or are complementary to policies designed to curtail immigration. In more sense than one, despite there being no "smoking gun" to prove the link between one policy and the other, there are good bases to suspect that they are complementary policies that could be defined as reactions to demographic pressures from "aliens." The ensuing question, thus, is how these issues are interconnected and how this is related to the "genetic turn" in modern democratic nations. This is indeed what this book is about.

By analyzing how eugenic policies were implemented by social democratic and labor movements, and by countries such as Finland that could be considered as welfare social democratic (despite the fact that social democracy was not the leading agent of the welfare system), and searching the connection of eugenics and welfare productionism in the five countries we deal with, we question whether there is a continuing line connecting eugenics in the past with deep and benign ART programs in the present.

As noted, several observers will find this argument hard to swallow but, as we shall try to prove, it is the underlying ideology behind the new genetic technologies used in Western societies. More than ever, they rely on productivist rather than race considerations; however, the coincidence between the productive and the ethnic is the underlying characteristic of these procedures. In more senses than one, as we shall try to prove, the new genetic developments coincide with the shift in eugenic practices from the Nazi racial hygiene to a productivist one.

In our historical theoretical analysis we foster enlightenment on this point. Indeed, the idea of "racial hygiene," which maintained that biological differences among races would inevitably lead to national degeneration, was connected to Nazi racial politics. However, we tend to overlook the fact that democratic countries have also made use of eugenic politics in order to enhance their national stock (see, for example, Balibar 1991; Goldberg 1997; Miles 1989).

Of special interest in this book is the role played by social democratic and labor movements in the genesis of welfare productionism, and its connection with eugenic policies. The reason for that special interest is the fact that social democratic movements were supposed to present not only an economic response, but also a moral and democratic alternative to liberal capitalism (e.g., Kautto *et al.* 2001; Kosonen 2001; Nygård 2006; Taylor-Gooby 2004).

However, in this book we challenge this view and emphasize the social democratic or socialist productionist bias towards eugenics. In other words, we emphasize the connecting points between social democracy, national productionism, and eugenics, while our underlying claim is that since most European nations are national communitarian and democratic, their productivist capacities are dependent on what eugenics can offer.

To proceed in developing our central point we shall focus on four examples, each with its own merits, and set the parameters of the synthesis of national productionism, welfare, and eugenic politics. These are the examples provided by Swedish and Danish social democracy, the Finnish nationalist welfare state, and the Zionist Labor movements in the early moments of the state.

At first glance there is no way to compare the Scandinavian countries with Israel. While Zionist demographic concerns and Zionism itself are controversial —a national liberation movement to its supporters and a racist colonialist movement to its detractors—the Swedish and Danish welfare states are not seen as charged with ethnic conflict, and in Finland what started as a conflict between a Swedish-speaking minority and Finnish majority evolved into an inclusive nationalism.

Most of these societies, however (partially the Finnish), used labor movements to champion similar visions of a people's-home state serving to protect and enhance the national stock. All of these societies elaborated their welfare systems based on national productionism rather than universal rights. All of them believed in the central role of the state in economic planning and social engineering. In other words, organized labor and socialism were dramatically tied, allowing for a near synthesis between socialism and state nationalism, which is the basis of a productivist ideology. The connection between eugenics, productivism, and welfare, thus, is framed around several forms of concern about population in the four countries: sterilization, restricted immigration, and reproduction. In other words, labor movements in Sweden, Denmark, Israel, and Finland, in a lesser manner, held firm to the notion that welfare depends on productivity, which in turn depends on developing the right human stock.

Finally, as noted, despite nationalistic productionist tendencies, these countries differed from Nazi "national socialism" in that they did not exhort racial hatred, they were democratic, and they were relatively tolerant towards minorities (Castles 2004: 4–5; Kautto *et al.* 2001; Korpi 2003: 590; Nygård 2006). In sum, we claim that social democrats in Sweden and Denmark, and Laborists in Israel and Finland, all favored both democratic welfare policies' productivist nationalism, and benefited greatly from the ideology behind eugenics. In Finland, however, Social Democrats were not the leading agent of the country's turn to eugenics; nonetheless, Finland could be defined as a Lutheran social democracy, whose welfare-productivist values were shared by the entire Swedish-speaking minority and the Finnish majority as well as by the defeated Social Democrats and the victorious Whites in the Civil War. All these countries adopted the criteria that a welfare society was sustainable only through increasing the number of productive individuals while reducing the number of feeble individuals. Thus, they employed eugenics under the guise of "productive welfarism" with the hope of creating a state consisting only of those who positively contributed to the nation. Our concluding remark at this point is that the productionist basis of eugenics/genetics overlaps the defense of ethnic makeup. That means that the productionist definition provides the aura of universalism to an "exclusionist" definition.

However, this book would not have shed light on an interesting story of far away long past democratic societies' use of eugenic policies, if it was not that some current developments raise suspicions that the ideology and the rationale for eugenic practices have reappeared guised under the new trend of genetics. In this book we handle precisely this point. The examples of Sweden, Denmark, Finland, and Israel, despite their different socio-economic development, constitute different faces of the model of nation-state countries that pay special attention to the productivist face of their makeup. However, the important point is that this study brings an original focus to similar circumstances that set fertile soil for the renewal of eugenic practices worldwide. We maintain that the current discussion of new genetic practices as cost-saving of public health resources is directly influenced by eugenic ideology. We hint that developed welfare nations, challenged by economic competition and the pressure of immigration, might aim to reduce the costs of welfare dependency by using genetics to ensure that as many children as possible are born with genes that enable people to live full and independent lives with no need for welfare dependency (Fitzpatrick 2001; Willis 2005). With regard to the development of population policy in current times, we add to the four countries analyzed the Netherlands, Germany, and France. These three are also welfare societies, and have also adopted ART policies. However, what we want to stress is that the model of eugenic productionism set by the former four countries led to very high and benign ART programs, in contrast to France, the Netherlands, and Germany. The first four have followed different policies but converge towards the same population policy mix: different in the details of emphasis and structure, but similar in their national productivist practice. In other words, today's and tomorrow's population policies and regulation of genetic technologies in reproduction options are constrained by earlier developments. High levels of economic pressures and declining TFR (Total Fertility Rate) have paved the path for today's population policy. The longer a productionist nationalist path that is characteristic of a certain welfare state has been followed, the more difficult it becomes to leave this path (Grant *et al.* 2006; Sunde 2007; Bujard *et al.* 2012). At the same time, we suggest that the Nordic-Israel model is indirectly, and not assumingly, inspiring to other welfare democracies that want to protect their national culture, their welfare state, and their productive capacity.

In sum, as we prove in this book, the ideological subtext behind genetic research in Western democracies is related to a twofold strategy—productivity and protecting the national stock. Rather than opening the gates of immigration and implanting costly civic integration programs or universal welfare, the new trend in some Western European democracies is a marked preference for genetic research and eugenic implementation. The productivist underpinnings of these policies lead to the conclusion that the eugenic policies of the past have been refurbished to provide responses to similar logic. In books by Elof Axel Carlson, *The Unfit* (2001), and Richard Lynn, *Eugenics: A Reassessment* (2001), the authors consider whether anything could be rescued from the wreckage of the 1940s. Carlson adopts a moralist position, claiming that we should erase from our vocabulary the concept of the unfit. Lynn, in opposition, states that authoritarian countries such as China that pursue eugenic policies will dominate democracies that do not. His conclusion thus is to be aware and adopt eugenic policies. In this book we deal with eugenic and productivist democracies, precisely pointing out not only that the idea existed and was promoted, but furthermore that it is still being advanced through genetic research and development. While abstaining from presenting a moral claim, this book presents the existing options that without any doubt might be enticing to current welfare democracies.

The book's outline

The book comprises five chapters.

Chapter 1: Eugenics and productive socialism

We begin our study by briefly analyzing the intellectual roots of productivist socialism and its links with eugenics. The first part of the chapter focuses on a brief history of eugenics, through which we attempt to uncover an ethical dimension to decisions about the application of scientific knowledge and technology in welfarist systems. We attempt to address how eugenics was being incorporated into population-based analysis and became politically prominent in various political settings. Thus, in this chapter we claim that population policies are designed in a context where citizens and government agencies share a commitment to welfare-state policy choices, thereby limiting the range of other possible trajectories. Within the context of national welfarist productivism, we explain how eugenic ideas took a particular form concerning subsidizing population and fertility.

In order to deal with this last topic we address the issue of productive socialism and its relationship with fascist and racist nationalism, but especially with democratic socialism. There is no question that productivist socialists generally associate the productive sectors of society with the "dominant" ethnic group, and with this sense they become associated with nationalism. Indeed, when nationalism becomes the main engine of productionist socialism, eugenics comes to the aid of the national community in its aim to reproduce itself. From the last third of the nineteenth century until the late 1920s, race was a central category in eugenic discourse. Race in eugenic thinking included three main presumptions: The first is that there are essential biological differences among races. The second is that physical proximity of disparate races leads inevitably to both racial and social degeneration. Finally there is a political conclusion: social programs should be designed to stave off racial degeneration and competition. At the turn of the twentieth century, eugenics' outlines became central to the interlaced constructions of "race" and "nation." Races were intended to denote inherent biological classifications of people; nations, conversely, are presumed to be natural spatial divisions, often defined by racial homogeneity. A large part of metropolitan and colonial projects of nation building turned scholars' observations to the eugenic discourse of race and, as a result, the exclusion of those racialized as outsiders has been an intrinsic feature of nationalism (see, for example, Balibar 1991; Goldberg 1997; Miles 1989). It is not questioned that eugenics fit nationalism, especially racist organic nationalism that appeared at the end of the nineteenth century. However, as noted, despite the fact that the very idea that productivist socialism is related to the Sorelean syndicalist revision of Marxism, giving rise to fascism, we suggest that it is part of the democratic socialist traditions. At the intellectual level, the democratic face of the productionist socialist tradition was expressed by the Fabians. Led by Sydney and Beatrice Webb as well as George Bernard Shaw, the Fabians were the ideologues of British trade unionism. They have set the underpinnings of the British modern welfare state and portrayed the growth of joint stock companies as part of the raising of the productive side of society and, as such, as a step in the direction of socialism. In more senses than one, the Fabians promoted a model of industrial democracy, composed of "the healthy and fittest," setting the model of a eugenic productivist welfare society. Influenced by social Darwinism, their model of society certainly was not a liberal one. However, they rather idealized a democratic society in which inclusion was conditioned to productionism.

Chapter 2: From "race hygiene" to "national-productivist hygiene"

This chapter explores the nature of the eugenic ideas in Sweden, Denmark, Finland, and Israel before World War I, emphasizing the way race biology dominated genetic science and eugenic rhetoric in Sweden before World War I (Broberg and Tydén 1996).[8] The ideology of eugenics in Sweden was sustained by a long intellectual heritage of physical anthropology. However, in this chapter we stress the role of productivist social democracy in setting the stage for the synthesis between eugenics and the welfare state. From the early twentieth century, Social Democrats such as Alfred Petrén, a member of parliament and head inspector of all mental care institutions, defended what was considered to be the quality of the Swedish population. Petrén's initiatives signified a gradual shift of reasoning about and managing sterilization of the mentally deficient from racial categorization to social-productive, which fit social democratic goals of defining the boundaries of a Swedish people's home. Swedish social democracy promoted a middle road between German social democracy, which gave priority to theory over practice, and the English absence of theory (Berman 1998). However, despite the low ideological level of Swedish social democracy, it still promoted two important principles—political democracy based on the right and obligation to universal suffrage, and social democracy based on the central role of the unions. As noted, the role of the state on the one hand and the unions on the other sets the classical formula of productivist socialism. This was better represented in the manifesto that could be defined as the most authentic manifesto of the Swedish welfare state: *The Crisis in the Population Question* written by Alva and Gunnar Myrdal (1935). In this chapter we analyze the Myrdals' opposition to immigrants, which could hardly be included in the nation, and we shall scrutinize their claim that decreasing breeding figures had to be actively combated with positive welfare measures.

This ideological background led to the debates on sterilization in 1934, which were directed to the mentally ill, and in 1941 were extended to allow sterilization to combat "antisocial behaviour." In more senses than one, the Swedish type of eugenics represents the clearest rationale of a socialist productivist stance, totally connected to the ethnic character of Swedish national culture.

Investigation of the nascent eugenics ideas during the 1930–1940s in Israel provides an unusual perspective from which to analyze the project of social engineering that underlay the campaign to settle Jews on the land. Despite the particularity of the Israel Zionist example, which observers may argue could not serve as a model, we suggest that it fits well into the debate of national makeup by integration of its "own ethnocultural" Diaspora coming from abroad, while limiting the growth of the "non-wanted" ethnic minority.

The population debate in Israel thus was shaped by the unique realities of Jewish history. The Israeli Zionist movement was committed to transforming the Jews of the Diaspora, who lacked a cultural bond with any territory, and replanting them in a populated territory that others controlled. Zionism thus entered into an inevitable conflict with the Palestinian national movement. The Zionists did not display a hateful racist approach towards Arabs, believing instead that a developed Zionist country would benefit Arabs as well. However, despite their intentions, defining Zionism in ethno-national terms prevented any possibility of communal existence for the mutual benefit of Jews and Palestinians (Peled 1992). In its early days eugenic thought in Israel had a clear productionist flavor. During the 1940s and especially with the emergence of the State of Israel, a new mass immigration to Israel, composed of Middle Eastern Jews, began to raise eugenic concerns about the makeup of Israeli citizenry. The goal of creating a "new Hebrew man"—a fighter and a worker—demanded eugenic practices. Although the Zionist solutions luckily did not come close to those of the Swedish sterilization policies, managing the Jewish mentally ill in Palestine broadly fit into the eugenic framework of preventing the "problem of unproductive people" through immigration policy or eliminating it by means of asylum or repatriation. These solutions were complicated by a lack of medical facilities, limited funds, the presence of an imperial power, and the internal political conflict continually brewing in the land. However, in both Sweden and Israel eugenics was promoted by productivist labor movements. In both countries, health practitioners and scientists posited a common ideology of productivism and welfare, integrated with a national concern that the population whose members were mentally or physically sick were profoundly damaging to the survival of the nation state. Yet observers have good reason to suspect that the synthesis of welfare productivism and eugenics aids those who want to protect the nation by excluding those that are ethnically different.

Israel's demographic discourse became shaped by a preoccupation with ethnicity and nationality. Demographic research in the 1970s highlighted Israel's unique population composition and the sinking birth rate of the majority population, secular Jews. As we shall try to explain, a ramified system of endowments are provided to enhance breeding of Jewish secular "productive" families. A question that we shall try to answer is whether and how Arab productive women are discriminated against, while non-productive ultra-orthodox Jews are included within the parameters of a universal productivist ideology. The contribution to the nation's spiritual stance is included within the definition of productive work. While during the early years of the state the eugenic parameters were destined to differentiate between productive and non-productive Jews, in the last decade, as a result of the demographic hysteria, the national-communitarian part of the eugenic formula has been emphasized.

The population debate in Denmark is also marked by its productivist characteristics. Slightly different from other Nordic countries, Danish nationalism is historically and culturally grounded. Whereas Swedish nationalism is linked to the dramatic development of the economy after World War II, in Denmark we can trace the development of cultural nationalism, which precedes the development of a strong economy.

Social democracy gained power in Denmark without major conflicts with traditional parties of the center and right. This setting was the underlying basis allowing for enacting a number of reform laws that gave rise to the Danish welfare state. Precisely in order to preserve that kind of social harmony, Denmark was the first Nordic country to introduce a national legislation on eugenic sterilization.

The early origins of eugenics in Denmark could be traced to the widespread belief, since the nineteenth century, in degeneration. In the early twentieth century, eugenics was perceived as a response to hereditary determinism, which focused on what was considered social ills. However, with time, the progressive side of eugenics was developed by leftists such as Gustav Bang and Wilhelm Johannsen, one of the principal architects of the new Mendelian genetics in Denmark. Very soon the adoption of sterilization methods based on progressive thought was adopted by social democrat administrators influenced by progressive intellectuals. The central belief was that in order to survive, any welfare society should do its best to reduce its social margins. In this chapter we stress that the productivist approach best suited Danish social democrats. K.K. Steincke, a social democratic minister of Health and Welfare and Secretary of

Justice in the 1920s, was a clear representative of the new type of social democratic administrators.

Steincke advanced an approach characterized by a seemingly difficult synthesis of health policies and eugenics. The main idea was to protect and feed the feeble as long as they did not reproduce, while at the same time the state should encourage the most intelligent part of the population to be as prolific as reasonably possible.

In this chapter we stress the negative aspects of eugenics. The commission nominated by Steincke in order to "consider policies of sterilization of mentally retarded and sexual offenders" came with the report *Social measures toward degeneratively predisposed individuals*. Although the question of the voluntary character of sterilization was enhanced, in reality there were no major differences between the laws of sterilization in Denmark and those enacted in Nazi Germany.

How do we make sense of this situation, if we claim that Danish laws of sterilization were not racist? In this chapter we explain that the Danish consensus was that eugenic measures, even those of the Nazis, were productivist not racist. Eugenics was portrayed as a neutral technical field of social policy that could be used by reactionary racists as well as by progressives. For all purposes, the Danish criteria of nationhood and eugenics were "productivist" rather than racist, while the Nazi sense of productivity was demolished by the priority of racism. In other words, the eugenics-based productivist-welfarism coincided with the understanding of Danish identity as "exclusionarily" inclusive—a welfare society for the healthy working class regardless of religion and origins. However, the social margins were to be reduced. From 1929, the year that the first laws of sterilization were enacted, to 1935, 108 people were sterilized; in the next five years the number rose to 1380. During the Nazi occupation, and regardless of it, the number came to 2120. Most of the women and mentally retarded were sterilized without consent. During the occupation the Danish population protected its Jewish members. Nonetheless, eugenic sterilization of socially marginal citizens continued and increased during that period.

The case of Finland illustrates the development of a welfare productionist society in which the Social Democrats did not have a leading role, although they were important in promoting the sterilization laws of 1935. Interestingly, despite not being led by social democracy, Finland is considered predominantly a Lutheran social democracy, which after independence and civil war decided to become a modern European welfare society. More interesting, the very idea of promoting a welfare society was adopted by the White bourgeois victors of the civil war that defeated the social democratic Reds.

The victory of the Whites redoubled their intention to achieve national unity and national integration of all parts of the citizenry, which was not only divided ideologically but also ethnically, between a minority of Swedish-speaking Finns and a majority that spoke a wide variety of languages of Northern Russian origin and even Mongolian. At first, eugenics was a mechanism of the Swedish minority to "stay alive" and dominant. They considered themselves modern

Europeans who dominated public administration since Finland was an autonomous Grand Duchy of the Russian Empire. Indeed, if we look for the sociopolitical background, we can say that if the Fabians in Britain projected their eugenic trend on a social class basis, in Finland the idea of racial hygiene was first embraced by the Swedish minority.

With independence there were various attempts to craft a Finnish identity with a strong sense of national purpose. As could be expected, eugenics became an important tool in the process of crafting national unity under a welfare-productivist society in which the power of professionals is enhanced. People trusted experts and professionals, and medico-legal experts were especially esteemed.

In this chapter we stress that since the 1920s, the large Finnish population adopted the language of eugenics. We emphasize that both communities, the Swedish-speaking and the Finns, began to share a common technocratic "cultural identity" and became worried about both the decreasing population of Finland and its qualitative deterioration.

During the 1930s, the increase in criminality was explained as a problem produced by the number of unmarried and the number of able bodies receiving poor relief treatment. These trends were considered contrary to productivity.

The debate focused on preventing people with unwanted hereditary characteristics from procreating. The debate in the parliament led to the Government Act of 1935, which allowed the voluntary sterilization of legally competent persons if there was justified fear that they would produce inferior children or if because of their sexual behavior they would be prone to abuse others.

At the same time positive eugenics was promoted. A good citizen was defined as a sober, honest, and hard-working member of the state. Motherhood and motherliness were self-evident characteristics of an ideal female. The upper classes, the holders of reform discourse, defined themselves as shepherds and guides of the common people.

There is no question that the Finnish spirit of those days, in which state propaganda was directed to mothers to give birth to more than four children, is continued in present days guised under a different title.

Chapter 3: The bioethical turn to biopolitics: the emergence of bio-nationalism

This chapter connects the history of productionist eugenics with the new research on genetics. It analyzes the revival of critical interest in eugenics in bioethics discourse regarding the new genetic technologies. The bioethics discourse addresses the question of whether there is a fundamental rupture between past eugenics and contemporary genetics, since the latter relies on consensus and choice. Thus, the bioethics discourse draws on dominant positions on life within traditional moral philosophy and the problems that may arise connected to the principles of justice, autonomy, and human dignity. However, in this book we are far more troubled by who has the power to control or even own "life," that is, the politics of life management. The co-option of eugenics as the bioethical discourse of the new genetics is intimately connected with what could be defined as the new interest of politics in population studies. This new stress on biopolitics is concerned with controlling population, rather than individual lives. It considers the transformation of individual lives into a calculable biological entity characterized as a "population," and of the inclusion of this emergent population at the center of political power. For that, the present book treats the normative discourse as a device to legitimize state control over diverse actors both internally (citizens) and externally (states). The nature of this interference differs depending on the governance system of genetic technologies. The policies of genetic technologies are not merely rhetorical exercises: they try to convince the population as a whole that certain policies are good for them, not merely for specific elite. Viewed in this way, they make the marginal population comply rather than rebel. They create normative bonds between the marginal and the rest of the population. These policies are also important for the formation of the national identity. They help states to define their vision of the world and their own role in it. Such policies always have ethical connotations and thereby cannot be seen as products of rational calculations only. But these policies also reflect complex historical and ideological processes that do not correspond neatly with rigid moral positions. The success or failure of genetic technology policies therefore depends on their ability to generate internal and external legitimacy rather than on their ability to meet moral criteria.

We consider here Foucault's framework of biopolitics with specific reference to eugenics. In terms of biopolitics, eugenics proves to offer useful power/knowledge tools for the rational planning of the population in welfare nations as it constitutes "positive" techniques to improve population life, while excluding "degenerate and unproductive" groups within a given population. Thus this chapter provides an analytical framework to organize comparative analysis of various national population policies and to assess their convergence into eugenic-based welfare population policy. Three major dimensions of positive eugenic population policy are identified, including immigration, family planning, and ART. Foucault always insisted that one of the signs that power was being exercised was resistance to it. If eugenics was the policy of democratic governments, therefore, one would expect to see challenges in the press and from various social groups, perhaps even opposition parties. One of the reasons that there was no cry of resistance is because these policies appear as progressive in favor of homosexuals, lesbians, single mothers, etc. The goal of this book is to shed light on the "exclusionist characteristics" of the ART processes that are hidden from public consciousness.

Chapter 4: The bio-regulation return of eugenics for a new century of massive immigration and multiculturalism

This chapter puts forward the claim that the use of eugenic policies to promote a welfare-productivist society is not dead. In this chapter, we explore the present use of eugenics to further democratic ideals, and argue that eugenics in current health care provision is linked with the notion of welfare-state growth. We start by exploring the role of "eugenic productivism" as a theoretical framework that extends the existing typology of welfare regimes to include the concept of people's nations' welfarism. Studies on types of welfare regimes provide ideal types of welfare regimes on the basic tenet of universally based mechanisms. The ultimate goal of these typologies is to measure the degree to which a country's welfare regime is universal, earnings-related, or generous. The major welfare-state regimes are namely "social democratic welfare regimes" such as those found in Finland, Denmark, and Sweden, which represent an inclusive, liberal, and generous welfare regime; "conservative welfare regimes" such as the type found in Germany, France, and Switzerland; and "liberal welfare regimes" of the type found in the UK, USA, Canada, and Australia.

This book argues that when explored at the national level, welfare programs are more likely to impose various forms of exclusionary practices in which different social groups benefit from social transfers, given their working status and citizenship. The possibility of an evolutionary convergence of welfare regimes, as suggested in this book, can be examined within the context of similar economic and political constraints faced by welfarist nations. Thus, this chapter offers brief economic and political overviews of country-by-country profiles of demographic and social trends that shaped the development of their welfare systems. This brief overview proceeds by a comparative analysis of the assisted reproduction practices and regulation along with immigration and family allowance regulations in Sweden and Israel, focusing on the imprint of national and economic priorities on each nation's population policies. The deeper significance of the comparison between Sweden's and Israel's eugenic policies stems from the fact that it marks one of the clearest contemporary examples of how their biopolitical strategies are now gradually diffusing into other European welfarist countries.

Chapter 5: European countries and eugenic productivism gospel—potential convergence trends in eugenics productivist population policy

In this chapter we analyze the broad range of declining TFR pressures, focusing on policies targeted at women and couples including ART technologies, family planning, and immigration policy. We discuss the convergence in ART and family planning trends among welfare democracies such as Finland, Denmark, Germany, and France. Although national response to changing demography has varied with the institutional pattern of welfare regimes, despite budget constraints, national governments continue to fund genetic and family planning policies for workers and their families.

In sum, this book bridges the gap between eugenics and genetics through the concept of welfare productionism. By establishing the category of welfare productivism, akin to nation states that want to protect their national culture and national stock, this book permits the understanding of scientific genetic initiatives in current Western democracies as a continuation of the eugenic practices of the 1930s. By disentangling eugenics from its racist flavor and including it within the frame of production, we strive to shed light to the real meaning and substance of these practices.

While not necessarily racist, the eugenic flavor was akin to nationalism, and its productivist face highlights precisely its nationalist exclusionist, although not overtly racist, characteristic. The dominant ethnic group in Western societies is generally defined as the productive one, and through it the rest of society will benefit. In other words, this is neither about ethnic cleansing nor about any attempt to limit the nation to the dominant ethnic group. The main idea through the productivist initiative is that through the enhancement of the dominant group, through the preservation of its dominant majority, productivity and national identity are preserved. In sum, this trend could be defined as "exclusionary-inclusionist"—neither exclusionist racist nor inclusive liberal multicultural.

Without any doubt the subtext of this ideology runs deep inside the national psychology of Western European democracies in current times. Today more than ever, precisely because of globalization, the Israeli and Swedish research and operative use of genetic practices might be enticing to Western democracies. Since these eugenic practices may be the result of a cynical spirit, the best recommendation we can offer … is to define them with another name.

1 Eugenics and productive socialism

New genetics and eugenics—old wine in new bottles?

We believe that the history of eugenics is instructive for those concerned with the bioethics of the new genetics. We do not pretend to provide here a comprehensive bioethical theory that would connect ethical virtues to uses of new genetic technologies. Instead, our aim is to uncover an ethical dimension to decisions about the application of scientific knowledge and technology in welfarist systems. In particular, this book provides some essential materials for constructing a eugenic productivist framework for discourse about the ethical deliberations of the use of reproductive technology. Based on the twofold nature of eugenics as science and politics, two discourses conflated in examining the nature of eugenics: the scholarly discourse situating eugenics in the history of science and aiming at understanding the intellectual development in scientific and ideological contexts, and the political discourse that we set out in this book to compare the historical and the current conjunctures between eugenics and new genetics in population policies in welfare states.[1]

Thus, this chapter seeks to address how eugenics was being incorporated into population-based analysis and became politically prominent in various political settings.

In this book we claim that population policies are designed in a context where citizens and government agencies share a commitment to welfare-state policy choices, thereby limiting the range of other possible trajectories. Within the context of national welfarist productivism, the eugenic ideas take a particular form concerning subsidizing population and fertility.

Several historical studies of eugenics have emphasized that the discursive diffusion between science and politics has significance in explaining the rise and fall of eugenics. Eugenics as an applied science sought in its beginning to live by the doctrine of unity of theory and practice for improving society and to impose certain obligations for political action on its adherents. Most of the historical studies of the eugenics movement have proved in more ways than one that in its early beginnings, eugenics' ambitions were not much different from totalitarian ideologies such as Nazism, fascism, and Leninism in various national and cultural settings (Adams 1990).

However, the interesting thing, as Porter suggests, is that Galtonian ideals adapted not only to totalitarian ideologies but to new practices of welfare planning (Porter 1996). According to Kevles, since its beginnings in the 1860s, eugenics has undergone two major transitions: the first was the shift from "mainline" to "reform" eugenics and the second from reform eugenics to contemporary medical genetics (Kevles 1985). Mainline eugenics is defined as the product of the first meeting of Darwinism and medicine: the understanding that evolution is ongoing, and the emerging concern that the human evolutionary process be rationalized and guided by the ideal of progress, and the human stock protected from harm (Kevles 1985: 88, 173–176). At times Lamarckian (assuming the inheritance of acquired characteristics), mainline eugenics, however, failed to distinguish between cultural and biological heredity. It emphasized the determinant character of the latter (Gems 1999).

Advocates of mainline eugenics called for controlling the reproduction of inferior stocks, which could be defined as lower classes or foreigners.

From the 1920s, however, scientists began to understand that Darwin's theory of evolution was insufficient and was surpassed by a more rigorous understanding of human genetics. As a result of this new consensus, a new type of reform eugenics based in human genetics emerged, the prevention of specific genetic diseases becoming its central focus (Gems 1999). Still, as we shall try to show, there is a direct connection between mainline and reformist eugenics. The latter is connected to the idea of productivist socialism, which represented not only a middle road between Marxist socialism and liberalism, but also between racist nationalism and liberalism. As we shall see, fascist as well as labor and social democratic movements were productivist socialists who used eugenics to create a scientific society relying on the society's strongest sectors.

While fascist productivist socialism was authoritarian, social democratic and labor movements promoted an organized democratic society, equipped to meet the demands of the modern world. Despite differences, social democrats and fascists supported the idea of a national community. Differently from the Nazi concept of a closed exclusionist and racially superior community, however, social democrats endorse the idea of a democratic republic unified in productivist values. Inclusion is conditioned to productivity, and eugenics became a central tool in order to set the basis of a modern productive national welfare society. Eugenics, thus, was a tool of modernization.

Several books and essays written since the 1970s have thoroughly proven that eugenics flourished in Edwardian Britain as a modernizing ideology for the scientific management of health care problems by the middle class (MacKenzie 1976; Paul 1998; Searle 1976). Moreover, in Great Britain, mainline eugenics evaluated class differences rather than racial differences (Zweiniger-Bargielowska 2007). In Scandinavian countries and Germany, race dominated institutionalized scientific discourse and influenced nation building and social democratic politics (Broberg and Roll-Hansen 1996; Proctor 1988; Weingart 1999). In the United States, biological knowledge about inheritance served as a discriminating tool for social goals during progressive eras. The existence of racial and gender differences entered at the point where historical events and circumstances intersected with personal and familial experience (Dowbiggin 2002). Afro-Americans and women became a potential focus of interest for eugenic management, as these groups were at high risk of experiencing poverty and welfare dependence.[2] Moreover, as shown by Hansen and King (2001: 243), eugenic ideas were also translated in the United States into policies by political actors who possessed powerful institutional positions. Indeed, the United States found a unique convergence between eugenic ideas and strategic interests: motivated by political opportunism and (in some measure) racism.

In short, historians have shown that eugenics thinking, as an avenue for the application of science to social problems, became compatible with varied political views, namely a "scientific ideology" (Spektorowski and Mizrachi 2004). Consequently, popular eugenics movements succeeded in rapidly introducing eugenics ideas into public discourse prior to World War II.

In proceeding from our working hypothesis that contemporary genetic practices are embedded in eugenics, we need to explain how its core ideas and practices are still with us today and not confined to the Nazi era.

Thus, in the following we will attempt to portray and highlight eugenics' "cluster of ideas and activities that aimed at improving the quality of human race through the manipulation of its biological heredity" (Kevles 1992: 4).

We shall set the theoretical frame enabling understanding of why these eugenic ideas that fit the productive interests of socialists are reproduced in the logic of genetics in present times. In synthesis we will trace the links between eugenics and socialist productivism, and in order to do so we should start by introducing a brief historical analysis of social science disciplines through three forms of discourse that together span much of the disciplinary spectrum of eugenics.

Since eugenics was not so much a clear set of scientific principles as a "modern" way of formulating social problems in biological terms, and also as a political spurious set of beliefs between the two world wars, we should employ two models for the purpose of eugenics analysis: scientific and ideological.

As an applied science, which could be used to guide social policy and practice over a wide range of problems in health care and welfare, eugenics fit the strategic research site for the study of science–ideology interactions. Because these two types of disciplines are apparently quite different from one another, it seems risky to conflate them haphazardly. The apparent incompatibility of science and ideology helps to provide a disciplinary account of the survival of eugenics' ideological core in social democratic movements.

Eugenics in accordance with a scientific model may not necessarily rely on solid rigorous research to back it up, because research in modern genetics and anthropology rendered eugenics' scientific authority senseless. A scientific approach often fails to take into account that the goals of eugenics are intimately tied up with the values of nation or community and that these values are often driven by ideology. Therefore, an eclectic model that connects the scientific research to a larger ideological context would be a proper way to investigate the continuity or discontinuity with historical disciplinary analysis.

In this chapter we argue that, in time, eugenics' ambition to reconstruct society was suppressed by social science's twentieth century core disciplines. Therefore, the history of eugenics should be connected not only to political history but also to the history of social science. Despite the evident variety of eugenic ideas, three attributes of eugenics have been alleged to be inherent in the core eugenic doctrine of improving the stock of humankind by application of the science of human heredity: race, artificial selection, and degeneration. These attributes were constitutive discourses of scientific disciplines, namely anthropology, human genetics, and degeneration. These three discourses will be used to trace the disciplinary status of eugenics to account for paradigmatic shifts in anthropology, human genetics, and ideology prior to World War II.

First, we note the failure of eugenics to account for paradigmatic shifts in anthropology through the conceptualization of race. The category of race will allow us to determine the paradigmatic status of anthropology as a discipline. Within the scientific model, eugenics succeeded or failed to account for theoretical shifts in the racial category within anthropology. Of course, anthropology succeeded in becoming a reputable discipline and eugenics did not.

Second, we examine how eugenics had been interwoven by the discourse of artificial selection into the discipline of human genetics. It is suggested that eugenics refused to regard the model of variability in large populations as compatible with new human genetic heredity. When the modern synthesis established itself as a permanent feature of the genetics landscape, eugenics could no longer provide a conceptual territory within the biological sciences in which the discipline could operate. Finally, during the 1930s there was considerable agreement among geneticists about what constituted the paradigmatic core of a genetic discipline, namely a revived Darwinism as the dominant theory of evolution. Consequently, the synthesis form of Darwinism led to an interpretation of evolution in terms of the genetics of populations, new factors being created by mutation and established in the population by natural selection when they conferred adaptive benefit (Mayr 1991). A modern synthesis was established as a permanent feature of the genetic scientific landscape.

Third, we will shift into the ideological political realm and explore the ideological construction of degeneration over time in both right and left political wings while adapting to changing political realities. Within the ideological discourse, we examine the idea of degeneration as a core concept in several ideologies, including eugenics, with attention to the interplay of its respective (peripheral) loosening and tightening ideological forces, namely, class, youth, heredity, and race. We will show how the interaction of these features found a point of inconsistency in the eugenic ideology of degeneration.

When analyzing eugenics within the framework of these models, we can locate just how eugenics represents a major overhaul of the nation's welfare system. It is suggested that eugenics' absence of clearly and coherently defined scientific and ideological objectives make it possible for eugenics to play a role in the nation's welfare system. Rather than being well constructed, eugenics' core objectives were arrived at through broader political and social consensus following World War II. To do this, eugenics shifted its emphasis from what was referred to as "race hygiene" to the more consensual form of "national productivist (social) hygiene" during the late nineteenth and early twentieth centuries. This shift is grounded in the scientifically disciplinary history of eugenics in which eugenics was recognized as a significant interlocutor in various areas of thought and social life, especially in the first third of the twentieth century.[3] Welfare policies were formulated on the basis of the values of the community and what measures, based on rational knowledge, will bring those values to fruition. Efficacy rather than well-constructed ideology was more relevant to the narrow question of whether a stated policy achieved a predicted set of objectives.

Finally, the continuity of eugenics in contemporary human genetic practices should be based on rigorous scientific research, while also factoring in the values of the broader society, that is, ideology. This would imply that despite the fact that eugenics lost its scientific authority or its scientific reliability to measure whether the policy is achieving the stated goals, its survival derived from the relevance of its scientific vocabulary to social fears and social political panic. Its old prestige of science allowed social reformers to embrace its main tenets. Thus, in this chapter we present why eugenics became so important for socialism.

This purpose branches into a broader theoretical debate that comprises the succeeding three chapters: the present chapter defines the nature of eugenics, the succeeding chapters compare how eugenic ideas were translated into population policies and which form of eugenics accompanied population politics before World War II up to today in welfare states (Marshall and Rose 1989). It is suggested that the ideational features of eugenics thus securing a place for path dependency have been delineated as factors influencing welfarist policy and politics.

Eugenics between science and ideology—identifying the eugenic disciplinary core concepts: race, artificial selection, and degeneration

Introducing eugenics

Historically, the term eugenics has been defined as a science that investigates ways to improve the genetic conditions of the human race. In general terms, eugenics synthesizes science and policy, and as a discipline and social movement lies at the interface of biological science and society. Its most important and first characteristic is the conviction that human traits of character are genetically transmitted. This so-called science, therefore, aspires to regulate human procreation by encouraging the fecundity of allegedly genetically superior groups in society, while discouraging "defectives" from producing children, since they would replicate their deficiencies.

Although the term eugenics was coined by Francis Galton (1822–1911), the intellectual history of eugenics could be traced back to the philosophies of ancient Greece. In *The Republic*, Plato's thoughts on eugenics and the breeding of philosopher kings and guardians were stated: "Defective off-spring ... will be quietly and secretly disposed of" and that medicine provided by the state "will provide treatment for those ... citizens whose physical and psychological constitution is good; as for others, it will leave the unhealthy to die" (1974: 174). Eugenic thought was also discussed in Roman times; its practice was employed by the Spartans who used to cast out babies who were considered unfit, including female babies, in order to protect their "pure" stock.

In modern times, as noted, the term eugenics was popularized by Galton, Charles Darwin's cousin, the founder of the biometric movement in England. It enjoyed its greatest influence from approximately 1905 to 1930, but the movement fell into disrepute because its scientific structure was faulty.

In 1883, Galton's elaboration of eugenics referred to one "good in stock, hereditarily endowed with noble qualities" (1883: 24–25). Galton's American disciple Charles B. Davenport (1866–1944) defined eugenics as "the improvement of the human race by better breeding" (1910). In 1904, Galton founded the National Eugenics Laboratory, followed by the Eugenics Education Society in 1907, whose stated aim was to educate the British public about eugenics.

As an applied science, thus, the practice of eugenics referred to everything from prenatal care for mothers to forced sterilization and euthanasia. Galton divided the practice of eugenics into two types—positive and negative—both aimed at improving the human race through selective breeding. Positive eugenics attempted to encourage reproduction by persons presumed to have desirable inheritable traits, whereas negative eugenics aimed at discouraging reproduction by persons having traits that were life-threatening, harmful, or of no civic worth (Galton 1904).

We turn now to a disciplinary history of eugenics along three attributes that have been alleged to be inherent in the core eugenic doctrine of improving the stock of humankind by application of the science of human heredity: race, artificial selection, and degeneration. These attributes were constitutive discourses of scientific disciplines, namely anthropology, human genetics, and degeneration. These three discourses will be used to trace the disciplinary status of eugenics to account for paradigmatic shifts in anthropology, human genetics, and ideology prior to World War II.

Race within the anthropological disciplinary discourse

The link between physical anthropology and the early eugenics movement became prominent in the nineteenth century since both scientific disciplines were dominated by national concerns. However, on the purely academic level, anthropology is regarded as a social science discipline that serves the purpose of removing politics and ideology from scientific research, basing the process on rational methodology, and not on political ideology.

To make the scientific model work, one should supplement the scientific explanation of the disciplinary diffusion between eugenics and anthropology through their shared conceptual category—race.[4] In the eighteenth century, anthropology attempted to fit race discourse into a general theory of history and progress. Anthropologists commonly defined a human race in terms of physical and anthropometric characteristics. Europeans had believed in the biological reality of races for a long time, and in the taxonomic schemes of Linnaeus and other pioneer systematicists, they were given a status almost equivalent to that of a species. In the nineteenth century anthropology was almost wholly devoted to the comparative study of race, and the effort to fit it into the general evolutionary theory.[5]

Nineteenth century anthropology saw a further development of the concept of race that can be traced from the conflation of social and biological evolution first proposed by Spencer (1966). Here, to over-simplify, we suggest that Spencer combined Malthus, Lamarck, and aspects of Darwin to offer a synthetic understanding of the evolutionary mechanism necessary to explain "higher" evolution, especially the social development of humanity. Spencer believed in the Lamarckian idea of inheritance of acquired characteristics in order to secure sociological progress of adaptation to local environments and conditions. Progress was conceived in terms of population quantity and quality. Thus, Spencer accepted Malthus's formulation of the causal overpopulation that causes poverty—the poor are poor because they have too many children. They are, in short, responsible for their own condition.

Charles Darwin, writing some years later, was ambivalent over the perceived opposition between legitimate moral concerns for the weak and the need for natural selection. Darwin claimed that

> We civilized men … do our utmost to check the processes of elimination; we build asylums for the imbecile, the maimed, and the sick; we institute poor laws; and our medical men exert the utmost skill to save the life of everyone to the last moment.… Thus the weak members of civilized societies propagate their kind. No one who has attended to the breeding of domestic animals will doubt that this must be highly injurious to the race of men.
>
> (Darwin, 1874: 151–152)

Similar to Malthus, Spencer and his followers claimed that the natural forces of society should be left alone to discard its weaker members:

> Spurious philanthropists … to prevent present misery … entail greater misery on future generations blind to the fact that under the natural order of things society is constantly excreting its unhealthy, imbecile, slow, vacillating, faithless members, these unthinking, though well-meaning, men advocate an interference which not only stops the purifying process, but even increases the vitiation—absolutely encouraging the multiplication of

> the reckless and the incompetent ... and discourages the multiplication of the competent and the provident.
>
> (1966: 147–148)

Many anthropologists began to believe that races possess distinct mental as well as physical qualities and that certain races were innately "superior" (Knox 1850).

Once eugenicists coupled social progress with biological evolution, it became easier to explain cultural variations by suggesting that dark-skinned people had not progressed at the same rate, biologically or culturally, as had white-skinned people, and that this must be due to an inferior genetic endowment.[6] Francis Galton strove to advance a scientific discipline "of improving the human race" that would study "all the influences that tend in however remote degree to give the more suitable races or strains of blood a better chance of prevailing speedily over the less suitable than they otherwise would have had" (1883: 24–25).

Galton pressed inheritance into the service of mankind's attributes and potentialities in a discursive practice around the concept of race. The coupling of quantity and quality in issues of human race was the driving force behind the social, medical, and political discourse of birth rate at the turn of the century. As Galton himself put it, the less-eugenically desirable varieties were outbreeding "those whose race we especially want to have" (1883: 318). Galton introduced race as a significant variable in social progress.[7] "Those whose race we especially want to have, would leave few descendants, while those whose race we especially want to be quit of, would crowd the vacant space with progeny" (1883: 207).

Before the outbreak of World War I, anthropology was associated with an evolutionary approach concerning physical characteristics employed in anthropological classifications. Biometric techniques, used by anthropologists, challenged the doctrine that the physical type of any organism was determined by genetic components alone. Anthropological investigation implied that environment modified the physical features of migrated populations. This conceptual move was reflected in Haddon's *The Races of Man and Their Distribution* (1909). The book dealt with some of the physical characteristics employed in classification and the grouping of main stocks according to those characteristics together with their geographic distribution. According to Haddon, "Various methods are employed in the attempt to group together different human communities and to distinguish between the races of mankind: these may be briefly described as physical, cultural and linguistical" (1909: 1–2). Alternative terms for races were suggested, such as "people," to account for the variety of communities including those with a racial mixture.

Pre-World War I, American anthropology was associated with Franz Boas, who published *The Mind of Primitive Man* in 1911 (Boas 1938). The main thesis of his book was that race, culture, and language were not interacting variables and should not be confused. Migration and historical changes had caused the three categories to overlap, but in principle they remained separate. Boas argued,

"The biological unit … is much larger than the linguistic unit." The question of the "melting pot" was much debated. The debate surrounding the issue of the melting pot within the American context centered on the effects of immigration of millions of prospective new citizens to the United States. Boas's immigrant research results aimed to challenge the doctrine that the physical type of any organism is determined solely by hereditary or genetic components.

He questioned the assumption that environmental forces were unable to affect the end result of their functioning (Herskovits 1973: 4–5). In addition, Boas showed that the White claim of general racial superiority, particularly Anglo-Saxon superiority, had no scientific support. He recognized that the differences between races were small compared to the similarities, that there was more variation within a race than between races, and that the many intermediate degrees of variation between races made the distinctions between them arbitrary. Boas worked on physical measurements of various "races" in a series of investigations in which he argued that the environment modified the physical features of migrated populations.

During the period before and after World War I, anthropological discourse came to refute the very notion of physical stability, which was the basis of racial theory. One can conclude that these attempts to overcome the dichotomy of heredity versus environment represent anthropology's withdrawal from the twilight zone dominated by racial-genetic reductionism. Boas's study of immigrants' skulls undermined the traditional approach to physical anthropology (1940: 74–75). In his article "New Evidence in Regard to the Instability of Human Types" (1916), Boas insisted on the environment's influence on the differences in appearance between immigrants and their descendants. He pointed to the environmental differences between city and country to explain the results he had found (1916).

During the period preceding World War II, the anthropological community continued its intellectual struggle against racism, shifting the subject matter of its research from the biological to the cultural. The book *We Europeans* by Huxley, Haddon, and Carr-Saunders (1935) reflected a combined biological and anthropological effort to challenge the Nazi pseudo-scientific theories of race. *We Europeans* was primarily noted for its endeavor to combine biological and anthropological data in one volume in a serious application of advanced theories of genetics to a comprehensive study of ethnic groups.

It is interesting to note that in Hitler's book *Mein Kampf*, his "racial" characterizations and differentiations, more especially of the Jews, are based not on any biological concept of physical descent as to the essential nature and meaning of which he exhibits neither knowledge nor insight—but almost entirely on social and cultural elements (Huxley *et al.* 1935: 28).

It was suggested that the new terminology offered in the book was "an advance on Professor Boas's proposal to substitute 'population' for 'race'" (Barkan 1992: 296–302). In 1928, Boas's *Anthropology and Modern Life* was published. His discussion of eugenics stressed the need to distinguish between inborn and learned aptitudes and drives, and cautioned against attempting to solve what were essentially culturally determined problems by invoking mechanisms of biological control.

> Eugenics should, therefore, not be allowed to deceive us into the belief that we should try to raise a race [of] supermen, nor that it should be our aim to eliminate all suffering and pain … Eugenics is not a panacea that will cure human ills; it is rather a dangerous sword that may turn its edge against those who rely on its strength.
>
> (Boas 1928: 120–121)

This reorientation in the anthropological study of human races is but one aspect of a general paradigmatic shift within the discipline. Such shifts led to the recognition that eugenics was incompatible with the dominant race paradigm in anthropology, and this eventually undermined eugenics' central premises (Falaize 1925).

Artificial selection within human genetics' disciplinary discourse

This section describes the paradigmatic status of genetics at the beginning of the twentieth century. To appreciate the nature of early human genetics, it is essential to recognize the field's essential characteristic. The most persistent biological controversy during the early twentieth century between biomericans and Mendelians has led some critics to contend that eugenics was bound to fail to provide a hereditary foundation due to its inconsistency and fragmented approach to population genetics. Paul and Spencer contend that basically biometricism, represented by Pearson and Heron, had a potential role as a catalyst for needed changes in heredity theory in genetics (Paul and Spencer 1998). Biometricians did not accept that Mendelianism offered a full explanation of heredity, especially for continuous characteristics, admitting the efficacy of natural selection. The implication for eugenics policies was clear—preventing the procreation of those who carried a trait of certain illness could have a limited effect on the frequency of this trait in the next generation. Thus, biometric criticism of recessive inheritance was oriented towards the long run and could not govern eugenics policies of changes in breeding patterns (Mazumdar 1992).

During the first two decades of the twentieth century, human genetics was torn by the feud between Mendelians and biometricians.[8] The Mendelians adopted the generality of Mendel's laws to stress discontinuous variation by the particulate nature of inheritance. The leading Mendelians—Bateson, de Vries, and Johannsen—did not adopt natural selection. Instead, they ascribed evolutionary change to the pressure of mutation. However, the biometricians did not accept Mendel's laws. They pioneered the use of statistics for analyzing the variability of large populations and stressed the importance of continuous variation as the material of natural selection. Their commitment to Darwinism had led them to reject Bateson's Mendelian program. Biometricians followed Francis Galton and Karl Pearson's studies. Pearson's eugenics was based on the theoretical foundation that variation and selection could be quantitatively measured by focusing on a continuous variation of traits (Pearson 1900: 407–408).

Despite differences between the biometricians and Mendelians, during the 1920s foundations were laid for an eventual consensus. Johannsen's discovery of inherited and non-inherited variation clearly demonstrated the sensitivity of genes to environmental influences. It showed that development is determined not by heredity alone but by the interaction of heredity and environment. Johannsen's studies stripped the authority of the biometric paradigm in the field of heredity (Sapp 1987). Johannsen's theory helped to heal the breach between the biometricians and the Mendelians; thereafter the two approaches to genetics proceeded together (Wright 1960).

Karl Pearson had regarded his model of variability in large populations as fundamentally incompatible with Mendelian heredity. He refused to entertain any possibility of compromise along the lines suggested by the Mendelians' experimental approach. But his student, Ronald Aylmer Fisher, eventually realized that Mendelianism had come to stay and began to investigate how populations would behave if heredity were controlled by larger numbers of Mendelian factors with overlapping effects.

Fisher was able to show that natural selection would be effective in shifting the relative frequencies of genes in a population (Fisher 1930). Fisher assumed that the population has a complex genetic structure in which many genes affect each characteristic. Since mutated genes are often recessive, natural selection cannot eliminate unfavorable ones by simply killing off all individuals in whom the harmful characteristic is manifest. Selection can only work to reduce the frequency of unfavorable genes, thus balancing the tendency for mutation to produce a constant flow of such characteristics. A trait that begins to have a selective advantage—perhaps because of a change in the environment to which a population is exposed—will slowly but steadily increase its frequency. Fisher's efforts to apply statistical techniques to the genetics of large populations paved the way for a revitalized theory of natural selection, which became the basis of the modern "synthetic" approach to evolution.

Like Fisher, J.B.S. Haldane calculated the effects of selection acting upon single genes within a large population with random mating and perfect Mendelian segregation.[9] Haldane showed that so rapid a rate of change in the population implied a 50 percent greater production of offspring by those individuals expressing the favored gene. This was a far more intensive selective effect than Fisher had postulated. Fisher and Haldane saw each mutated gene as an independent new factor fed into the gene pool to await the effects of selection. In fact, geneticists were beginning to recognize that Mendelian factors are not always discrete units; they often interact in such a way that the effect of two genes together is not the same as the sum of each acting on its own (Mayr 1976: 307–328).

Karl Pearson's studies suffered greatly because of his insistence that Mendelianism was incorrect. Pearson's statistical research validated the theory that inheritance is very powerful, since it indicates that selective breeding is a definite measure of human mental improvement. Pearson concluded that generations of directed artificial selection would reduce the necessity of intervention since the population would continue to reproduce the desirable characteristics (1900: 405–410).[10] However, J.B.S. Haldane pointed out that human population is so complex that artificial selection had no realistic hope of eliminating undesirable genes within a few generations. By the 1920s, population genetics was beginning to show how difficult it would be for artificial selection to remove defective genes that had become established in the human gene pool (Bennett 1983; Box 1978; Pastore 1949).

Sewall Wright used the interpretation of a population's genetic variability to construct a more complex view of the evolutionary process. Wright assumed that a wild population seldom comprises a single, fluid gene pool. He argued that in many cases the species would be broken up into relatively isolated local strains, with only limited interbreeding. In such small groups inbreeding would create new gene-interaction systems that would be seized upon by natural selection when they happened to confer some advantage. By introducing the possibility that local isolation could subdivide the population, Wright produced a model of evolution that fitted more easily into the work of the field naturalists (1968: 431).

The stage was set for a synthesis that would establish a revived Darwinism as the dominant theory of evolution. The synthesis form of Darwinism led to an interpretation of evolution in terms of the genetics of populations, new factors being created by mutation and established in the population by natural selection when they conferred adaptive benefit (Mayr 1991). Modern synthesis was established as a permanent feature of the genetic scientific landscape.

Degeneration within the ideological disciplinary discourse

An ideological model acknowledges eugenics as social practices embedded in culture and power relationships. According to the ideological model, eugenics represents a socially situated practices approach driven by a particular ideological position that is created by some adjacent ideas and some peripheral ones (Freeden 1983). Following Freeden's analysis, different ideologies may be distinguished by their core ideas within the peripheral ideas environment. In this section we aim to show how eugenics' core ideas were shared by different ideologies from one political spectrum to another, but it was its ill-constructed peripheral ideologies that made it easy for eugenics to be adapted to the changing consensual and accepted views of society.

The ideological discourse about degeneration is central to the twentieth century. This study focuses and analyzes key elements in the ideological language of degeneration. Tracing the constitution of the language of degeneration within selective study acknowledges the ideological-political complexity of the idea of degeneration, its over-determination, and irreducibility to a single origin or term. In Foucault's terms, degeneration is discursively activated to produce categories and distinctions (1972a, in particular, Part 2 chapters 1–7; 1980, in particular, Part 2 chapter 1, Part 4, chapter 2).

Degeneration was never reduced to a fixed term or theory. Rather it was a shifting term, which facilitated discourses of crude differentiation: between the normal and the abnormal, the healthy and infectious, the "fit" and "unfit," the civilized and the primitive.[11] The development of the concept of degeneration can be traced to the medical theories of Benedict-Augustin Morel and Cesare Lombroso. In the 1850s, Morel proposed that environmental factors, such as alcohol consumption, could produce degenerative trends in individuals that would be inherited by their offspring. For Lombroso, degeneration was concerned with the pathology of crime and criminality. Degeneration in the second half of the nineteenth century was scientifically constructed as an essentially moral and social threat that was manifested in social pathologies, such as crime, suicide, alcoholism, and prostitution (Pick 1989).

If discourses of progress during the nineteenth century relied on a view of the healthy, rational nature of man, degeneration degraded the early twentieth century social body so that it was ultimately viewed as unhealthy, producing profound pathology, infectious offspring, and immoral relationships. In a discursive sense, both degeneration and progress during the late nineteenth century appealed to the authority of natural sciences, particularly to evolutionary theory.[12] Just as evolutionary theory provided a paradigm for the ideological language of progress, so eugenics provided a paradigm for degeneration. The scientific construction of degeneration was established in the writings of the representatives of the eugenics movement: Galton and Pearson. Galton was concerned with racial degeneration, believing in the innate superiority of the White races over all other human populations.[13] Pearson, who headed the Galton Eugenics Laboratory, suggested that direct selection would avert racial degeneration.[14] Since degeneration is a seminal concept in the twentieth century, we will trace its linguistic constitution prior to World War II. This study will draw attention to the continuities or discontinuities between successive forms of degeneration from the late nineteenth century to the outbreak of World War II. The concept of degeneration could provide a discursive tool claimed by both left and right ideologies due to the different ideological shape degeneration has assumed over time and in various political contexts. Thus, the present study offers not only an exposition of ideological contexts but also a sustained attempt to compare and contrast the language of degeneration in several different ideological contexts of debate.[15]

Degeneration and productive socialism. Fascism vs. national socialism

There is an intrinsic link between the ideological language of degeneration and socialism. Indeed several works have pinpointed the links between eugenics and socialism. Indeed what is clear is that from the turn of the century to the 1930s, eugenics was not necessarily related to the political Right. As several authors have pointed out, eugenics became a respectable ideology among a variety of Marxists. Several scholars have pointed out that eugenics fit the scientific and reformist spirit of the Progressive era featured by vast economic and social change (Allen 1989; Freeden 1979; Pickens 1968).

Democratic Socialists and Fabians such as J.B.S. Haldane, G.B. Shaw, Havelock Ellis, C.P. Snow, and Julian Huxley are some of the wide variety of left-wing thinkers who believed that there was a need for some form of a socialist social reengineering to overcome inequality. They perceived eugenics as a promising possibility for progress and human liberation (Freeden 1979; Paul 1984). The interesting thing, however, is that in most cases, they did not follow social Darwinist lines in order to endorse eugenics. As Diane Paul suggests, social Darwinism was associated, at least in Britain, with unrestricted laissez faire and emphasis on individual choice. Eugenics implied a social and often a state concern with reproduction (1998: 570).

However, we should first and foremost distinguish the type of socialist left we are talking about. We use the term productivist socialism in order to define one thread of socialism, which in both its democratic and non-democratic forms, departs from Marxism socialism. While the democratic face of productivist socialism is symbolized by Sweden's social democracy and British Fabianism, the non-democratic face of socialist productionism is represented by fascism. They share a common inspiration in technological advance and they support "social engineering and planning" that prevails over ideology. In short they overlapped in their productionist approach. However, they differ in their value of democracy. Fascists considered a technocratic authoritarian state as a substitute for democracy while Social Democrats in Sweden, and Fabians in Britain, considered an industrial society as the safeguard of democracy. They composed the democratic face of productivist socialism.

However, to be precise, for a wide variety of scholars the idea of productivist socialism is associated first and foremost with fascism, and specifically to the shift in socialist political thought from Marxist socialism to fascism and national socialism (Gregor 1979; Sternhell 1979; Sternhell *et al.* 1994, especially chapter 5).[16]

Intellectuals such as Georges Sorel in France and Arturo Labriola in Italy, among others, set the stage for the creation of a new type of productivist socialism, which contrasted with Marxist socialism. Thus, for several scholars this syndicalist revision of Marxism should be defined as the most dramatic revisions of Marxist political thought. This "new socialism" left aside the idea of class struggle in order to endorse and stress the revolutionary anti-bourgeois potential of national mobilization.

Several of the most important ideologues of productivist socialism, such as Henrik de Mann in Belgium, Marcel Déat in France, and Oswald Mosley in England, were precisely supporters of this different or anti-Marxian type of national socialists who had embraced productivist socialism, especially during the period between the world wars.

All of them in different ways and at different paces, searched for a type of corporatist organization of society in the name of a new, virile, and productive socialism, which they considered to be the socialism of the future. While old

Marxist socialism promoted dialectical materialism as a liberating science, the task of which was to eliminate the false consciousness created by objective science, productivist socialism uncritically adopted objective scientific developments while adopting ideas of social engineering. Whereas Marxists indicted capitalism as a system of exploitative property relations, productivist socialists focused on the control of credit and money as abusive. Productivist socialists, thus, did not see the confrontation of bourgeoisie and proletariat as a given and as an engine of historical materialism. They rather saw that one feature of productivist socialism was setting the productive and healthy segments of society—namely workers and productive capitalists—against the parasitic elements—namely corrupt political parties, financial capital, and feeble people. As Sternhell in most of his works remarks, this type of socialism fitted nationalism.

There is no question that both Mussolini and Hitler when they referred to socialism meant this type of nationalist productivist socialism that fitted productivist capitalism. Indeed Musssolini and Hitler courted business leaders. However, they did not play on Manchesterian terms. They did not praise financiers, and did not think that the country should be led by parasitarian capitalism but by productive capitalism related to engineers and captains of industry. Their approach was productivist, appealing to the new managerial elite and to the leaders of industrial corporations who promoted technological leadership free from non-productivist constraints. This productivist socialism set the underpinnings of a new national solidarity for the healthy parts of society. As Charles Maier suggests, "The state would rest not on the outmoded civic ideas of 1789 and subsequent liberalism, but on individuals' functional role in the economy" (Maier 1987). In more senses than one, it could be argued that the idea of "productivism" gave Mussolini the bridge by which he could cross from his earlier volatile socialism to productivist socialism during 1919–1922.[17]

Concepts such as these justified imperialism and a fundamental resistance to social democracy, although they allowed alliances with independent syndicalist leaders.

Was this productionist trend followed by Nazi Germany? In certain ways it was. Oswald Spengler's ideas on the connection between the Prussian character and socialism expressed in his *Prussiandom and Socialism* (1919), envisaged a collectivist system relying on national labor and the collaboration of the engineer with political leaders. The *Verein deutscher Ingenieure* remained dedicated to the notion of an apolitical technocracy, which, they believed, national socialism would help to achieve (see Spengler 1919).

Likewise, as Charles Maier reminds us, the movement for *Schänheit der Arbeit*, growing out of the *Neue Sachlichkeit* of the late 1920s, sought a sober modernism and technological aesthetic.

All these notions were embedded in ideas of industrial management that predated the Nazis. As an example, even the Nazis' resort to compulsory arbitration developed out of the Weimar arbitration courts that since 1930 had been turned into pro-managerial authorities (Maier 1987).

The important point to emphasize, however, is that all these trends developed and expanded under national socialism, which emphasized ideas of blood and race over those of productionism.

An important distinction persisted, however, between Italian and German corporatist programs. Italian theories urged corporative organization as a remedy for economic backwardness, as compensation for the crumbling of the Italian economy, and as a step toward industrial modernization. In Germany, with slight opposition, all supporters of corporatism strove to protect small business, handicraft, and the peasantry. Despite Nazi support for technocratic modernity, they wanted to shield *volkish* small business from the competitive and oligopolistic tendencies that were pressing them toward bankruptcy (Maier 1987).

Furthermore, it is important to remember that despite Nazi support of technocratic modernity, race and protecting the *volk* still trumped productivism. Indeed, "social protection" for Nazis meant protection of race, and of the *volk*, which is best represented by small producers.

In more senses than one, fascists were less confined to notions of race. The totalitarian technocratic state, at least until the adoption in 1938 of racist laws, relied on a national productivist view led by productivist elites. As we shall see, the idea of national degeneration was not related to race decadence because of mixtures with others, or to the non-purity of the nation. However, at the same time fascists believed that political liberalism and democracy led to degeneration.

National and socialist productionism vs. degeneration

From a theoretical perspective, we should stress that the direct connection between this national productivist ideology and the language of degeneration is well documented, especially in the political ideology of Georges Sorel. Though Sorel moved across the political spectrum, he was committed to the dominant philosophical idea that degeneration was ultimately and directly scientific. Sorel dismissed the nineteenth century "illusion of progress," and in *Reflections on Violence*, he formulated the idea that the "march towards deliverance … is narrowly conditioned … by a profound conviction of our natural weakness." Sorel criticized the bourgeoisie for their economic and moral decadence (1999: 70–79). The dominant scene of degeneration, suggested by Sorel, displayed signs of degeneration: excessive consumption of alcohol and women. "The union leaders spend a great deal of time combating drunkenness … with the help of constant surveillance by comrades it does not seem impossible to succeed" (Sorel 1976: 78–79, 91–92).

There was without any doubt a direct intellectual connection between Georges Sorel's revolutionary syndicalism and Benito Mussolini. Though a socialist, Mussolini consistently rejected egalitarianism. He believed that great historical events are promoted by a virile leadership. The masses were expected to progress to new historical eras; however, they could do so only if they were led by people who were more intelligent and daring than they. During his socialist period this elitism found Mussolini agreeing with Lenin's elitist revolutionary ideas. Even after he had abandoned the left, elitism remained a major principle of Mussolini's fascism.

Mussolini's "revision" of his socialism along nationalist and voluntary lines is exemplified in his article, "The War as a Revolutionary Event" (Sternhell *et al.* 1994, especially chapter 5). This article labeled anxieties about growing pacifism as "degenerationism."

> We now want to attach our destiny to these "dead" forces in the name of a "peace" which does not save us from the disasters of war today and will not save us from the infinitely greater ones tomorrow, and which in any case will not save us from shame and from universal derision of people who have lived through this great tragedy of history.
>
> (Mussolini 1914: 27)

Attention to Mussolini's conceptualization of degeneration reveals a capitulation to ideological language, which drives degenerationism into an expression of myth, rather than scientific truth. Ever since the appearance of Sorel's writings, for the ultra-right the myth of the nation had acquired positive connotations:

> Our myth is the nation, our myth is the greatness of the nation! And to this myth, this greatness, which we want to translate into a total reality, we subordinate everything else.
>
> (Mussolini 1917a: 44)

An emphasis on the state plays a pivotal role in fascist theory (Mussolini 1933: 161–164, 1935: 9–13).[18] The nation provides an intuitive understanding and picture of the essence of fascism: "When the sense of States declines and disintegrating, centrifugal pressures exerted by individuals or groups prevail, national societies are destined to decline" (Mussolini 1929: 65). In the particular historical situation of wars, namely of national degeneration, a distinct imaginative dimension allowed direct access to ideological discourse: blood. Mussolini's notion of blood became narrative encouragement to youth rather than the biological and hereditary essence of degeneration. This shift is well emphasized in the growing attention to the national, instead of scientific, language of degeneration discourse.[19]

Within the context of national degeneracy, the nation's adolescents were becoming the pronounced object of the medicalizing discourse of fitness.[20] Youth, after all, held the nation's future stock of racial health:

> The brutal and bloody apprenticeship of the trenches will mean something. It will mean more courage, more faith, more tenacity. The old parties, the old men who carry on with exploitation of the political Italy of tomorrow will be swept aside.
>
> (Mussolini 1917b: 29)

The turmoil that followed on the heels of defeat in World War I plunged Germany into a five-year period of economic and political chaos. In Munich at the war's end, Hitler became the leader of the National Socialist German (Nazi) Party. While scholars might depict the similarities between Mussolini and Hitler, there were sensible differences. Indeed, Mussolini emphasized the more abstract theory of the state, while Hitler incorporated racism into his ideology. Within the ideology of degeneration, it was imperative to establish a hierarchy of racial difference. "It was the consequence of a consciousness of right which was rooted solely in the conviction of the superiority and therefore of the right of the white race," said Hitler in a speech in Dusseldorf Industry Club in January 1932 (Baynes 1969: 792). In fact, racism was necessary because miscegenation and race mingling would inevitably lead to the degeneration of Western civilization: "Historical experience offers countless proofs of this. It shows with terrifying clarity that in every mingling of Arian blood with that of lower peoples the result was the end of cultured people" (Hitler 1924: 286).

Hitler's racial ideas offered a serviceable narrative for the "otherness" of the threatening races: "Here again the result would have been catastrophic; a people of alien race expressing its alien ideas in the German language, compromising the lofty dignity of our own nationality by their own inferiority" (1924: 390). The emphasis on biology in a discussion of racial differences was encouraged by the fear of contagion and infection: "It must declare unfit for propagation all who are in any way visibly sick or who have inherited a disease and can therefore pass it on, and put this into actual practice" (1924: 404).

Unlike fascism, Nazi ideology gave the state only secondary importance. The state was only the arena in which the race built its strength and identified its leadership (Schiller 1925). In foreign affairs the state was the vehicle through which the superior race governed the inferior ones:[21] "Just as internationalism harms and thereby weakens the existing race value, and as democracy destroys the personality value, so pacifism paralyzes the natural strength of the self-preservation of peoples" (Hitler 1928: 32).

For Nazis, thus, the question of degeneration had a clear racist connotation. The Nazis turned the idea of productivism toward a concept of meritocracy based on racial terms. In more senses than one, a productionist socialist who could be a fascist or a social democrat could infer why Jews should be excluded if they are productive. This question leads us to examine the process of the shift from eugenics based on a meritocracy to racial eugenics in Nazi Germany (Spektorowski 2004; Spektorowski and Mizrachi 2004).

It is quite clear and well known that German eugenics preceded and developed independently of racism, and was not necessarily a prelude to the Holocaust (Weiss 1990: 42). While the main goal of eugenic scientists in the pre-Nazi era had been to formulate a strategy aimed at boosting national efficiency based on the parameters of a meritocracy, from 1933 onward eugenic scientists were identified with the claim that racial purity determined efficiency. Further studies were conducted in order to provide genealogies for individuals whose pure Aryan lineage was in question. In this context efficiency referred to a value of a general utility function of racial ideology. Eugenic scientists gained a pivotal role in composing racial testimonials and genealogies for the Ministry of the Interior after the passage of the Nuremberg Laws. They offered expert testimony in cases presented before the genetic health courts that eventually supported execution of the sterilization laws. It can be argued thus that with the Nazi seizure of power the idea of racial hygiene was totally implemented. Until 1933, not even Aryan sympathizers would have thought and defended the preservation or purification of Nordic racial characteristics as the focus of their attention.

After 1933, race hygiene combined the ideas of racial and genetic care. While the latter was in the tradition of the non-racist eugenics concerned with the management of the mental and physical traits of the population, the idea of race care was new. This does not mean that there was no continuity between the period prior to 1933 and that which followed.

For example, before the enactment of the sterilization law in 1933, many scientists were considered politically neutral from racist orientation, but this law set the course for a racist eugenics implemented by the Nazis. Although there were other legislative proposals, such as the Prussian proposal of 1932, the Nazi law advocated mandatory sterilization based on racial grounds.[22] The task of these courts was to provide moral and scientific backing for this project. The scientific role of scientists such as Lenz, Ploetz, and Rudin vis-à-vis the genetic health courts was to compose expert committees to "rubber stamp" government policies. In other words, it was clear that German eugenic scientists cooperated with the Third Reich in transforming eugenic policies into a centralized state ideology and in making it understood that race is something fixed, not subject to change. In more senses than one, this tradition focused on the improvement of one's own race, with the concepts of race and eugenics working in tandem (Bock 1986: 60).

Eugenics and productivist democratic socialism. The British Fabians on degeneration and productivism

Common sense observers would have no doubts in assuming that eugenics is associated with conservative and racist views. Conservatives, for example, would agree with Darwin's assertion in *The Descent of Man* (1874) that "We civilized men … do our utmost to check the process of elimination: we build asylums for the imbecile … we institute poor laws … thus the weak members of society propagate their kind" (Trombley 1988: 6).

Conservative political elitists would accept that the task of social policy should reverse this social decline, not through health or educational programs, but by preventing the birth of future inferior generations.

A complementary idea is promoted by racist theories, which agree with those eugenicists in identifying feeble people with a specific race or social group. However, the attractiveness of eugenics was not only shared by conservative and racist regimes. Democratic intellectuals, including among them socialists of the democratic school of socialism, also endorsed eugenic ideas.[23]

The question is why?

One of the explanations we suggest is that behind democratic socialism's attraction to eugenics was precisely its connection with the ideas of productionism and with the idea that a welfare-productivist society is possible only under the frame of the nation.

In this sense the productionist view of socialism, in either its democratic or authoritarian form, takes the concept of social reform as a technical rather than a political question, and promotes the central role of eugenics in social welfare (Spektorowski 2004: 105). To be precise, this version of socialist productivist eugenics does not say "first we need growth in order to pay for health care and other goodies." This type of productivist ideology suggests that welfare provision is a means to the end of economic progress. Institutionalized welfare provisions were not reserved for the neediest cases, but for strong productive workers in order to keep them productive. In other words, welfare should not be perceived as a right or entitlement but as an investment in manpower. The idea that the productive agents of society should reproduce themselves, while the social margins should be limited, is an integral part of this approach.

Eugenics as we can see is a direct companion of productivist socialism.

As we have already learned, the productivist side of socialism was especially connected to fascism and national socialism. However, productivist socialists were not only fascists. Fabian socialists and Swedish social democracy should be considered the democratic and progressive face of productivist socialists.

However, this should be taken with a cautious approach. Fabian socialists as well as the Swedish Social Democrats hardly represented the most democratic and universalist face of Marxist socialism. They revealed instead the democratic face of productivist socialism suiting the nation state and especially akin to use eugenic tools to enhance a productivist national democratic culture.

In total contrast to fascism, which rejected the value of democracy, the Fabians and Swedish social democrats were sincere socialists and sincere democrats. However, they asserted that only this productivist socialism could be presented as the basis for the protection of the democratic nation. In this sense they set an exclusionary basis for inclusion. The democratic nation belonged to the healthy productive people.

These productivist features became especially visible as the twentieth century approached. Indeed, early in the century a substantial although overlooked shift in eugenics presumptions from what was referred to as "race hygiene" to "social hygiene" began to take place. That was the time that the eugenics legacy shifted from focusing solely on heredity to considering other various factors, socio-economic as well as biological. This allows progressive socialists to appropriate eugenics for their own interests.[24] Indeed, socialist regimes also had to deal with economic efficiency and deliberations took place concerning whether a higher quality national stock might contribute to productivity. As could thus be expected, eugenic policies pursued in the name of social progress raise serious difficulties. The reason is that socialists claimed not only social and economic superiority over capitalist productivity, they also claimed moral superiority.

However, some of these gaps in socialist theory, between productivity and democracy and between democracy and eugenics, were bridged by the British Fabians.

The Fabian Society was a small but extremely influential group of British socialists. Its origins can be traced in the Fellowship of the New Life, founded in 1883, which looked for new ways to reform society in an ethical rather than in a political way. After breaking away from the Fellowship, intellectuals such as Frank Podmore and Edward Pease founded the Fabian Society, which was joined by G.B. Shaw and Sidney and Beatrice Webb, among other first rank intellectuals such as Edward Carpenter, Annie Besant, and Walter Crane. They contributed to the ideological lens of British trade unionism and actually are regarded as laying the foundations of the modern welfare state in Britain.

The society became famous with the publication of the Fabian Essays in 1889, through which the Fabian Society at first set out largely to present an alternative to the then dominant Marxist Social-Democratic Federation. The society's basic approach could be defined by reform from above, and in that sense Fabian socialism was and has remained essentially evolutionary and gradualist. The Fabians envisioned a nation managed by people much like themselves: middle-class professionals such as doctors, scientists, teachers, and social workers. Indeed, their ideal was a scientifically planned society that would empower experts rather than workers.

In contrast to Marxists, the Fabians considered themselves above class interests, and as noted by Eric Hobsbawm, they formed the basis of a scientific socialism, whose protagonists "would be 'middle class theorists' representing the interests of ... the trained, impartial, and scientific administrators" (Hobsbawm 1964). The Fabians believed that capitalism created an unjust society. The cornerstone of their economic theory relied on the Ricardian law of rent, which has proved that the wealth of landlords arose from their monopoly power over the soil. However, they developed their theory of rent to apply not only to land but also to capital and to personal ability. They considered large incomes to be chiefly rents arising from the possession of differential monopolies and maintained that these rents belonged properly not to the monopolists but to the community as a whole. Basic economic injustice could be solved through the socialization of monopoly incomes through social ownership of the monopolies. As a conclusion of this thought, it was clear that the ultimate end of the Fabian society should be to reconstruct "society in accordance with the highest moral possibilities."

However, more important than the question of social justice was the fact that Fabians believed that a capitalist society was an inefficient society. As an alternative to capitalist inefficiency, G.B. Shaw forecasted a modern developed and industrial society rooted in guild socialism. In his essay "The transition to Social Democracy," Shaw extols the Middle Ages' sense of order and harmony. In the English Middle Ages, he claimed:

> Every man had his class, and every class its duties. Payments and privileges are fixed by law and custom, sanctioned by the moral sense of the community,

> and revised by the light of that moral sense whenever the operation of supply and demand disturbs their adjustment. Liberty and Equality are unheard of; but so is Free Competition.
>
> (Shaw 1891: 232)

The reason for the collapse of this order, according to Shaw, is not related to a question of justice, but because its machinery was too primitive and its administration too naïve. Thus, it is not strange that based on the guild experience, what Shaw projects is a new modern guild society, a functional democracy, based in a professional strong administration.

According to Shaw:

> Guild Socialism, in spite of its engaging medieval name, means nothing more picturesque than a claim that under Socialism each industry shall be controlled by its own operators, as the professions are to-day. This by itself would not imply Socialism at all: it would be merely a revival of the medieval guild, or a fresh attempt at the now exploded self-governing workshop of the primitive co-operators. Guild Socialism, with the emphasis on the Socialism, implies that the industries, however completely they may be controlled by their separate staffs, must pool their products.
>
> (Shaw 1904)

This was a clear productionist approach that stresses technological professional leadership. However, the dark side of the "production problem," as common sense of course, was the "social problem group," the antithesis of the professional middle class, in that they were both inherently degenerate and highly fertile. Although well aware of the economic origins of social problems, the Fabians did not generally look to resource allocation as a means of explaining the existence of a social problem group. These convictions turned the leading Fabian intellectuals into clear advocates of planned society, and therefore they were clearly attracted by genetic planning (Trombley 1988: 34). There is no question that the whole idea of genetic planning and/or a society led by experts would lead to eugenic politics.

When the biological and the social became intertwined and the distinction between them blurred, a new scientific legitimacy was given to social exclusionist politics. Indeed, when eugenics became the basis of social engineering, it broadened the range of those destined to be excluded. This was the basis of the new socialism. The Fabians attacked old socialism because it rested on ideas such as the right to live and the right to work, "ideas of retrogressive rather than progressive selection" (Ball 1896, cited by Trombley 1988: 37). However, the new socialism was defined by one of the most famous Fabians, H.G. Wells, who argued in *A Modern Utopia* (1908) that in an ideal socialist society the state would forbid the procreation of those judged to be below the national minimum of "physical efficiency." Wells argued for the sterilization of failures on the grounds that the way of Nature has always been to slay the hindmost, and there is still no other way, unless we can prevent those who would become the hindmost being born (Wells 1908: 60). Also, George Bernard Shaw believed that there is no reasonable excuse for refusing to accept that only a type of eugenic religion can save Western civilization. Shaw added the "positive" side, and proposed a system of breeding outside marriage.

Thus, it cannot sound strange that Sidney Webb, who became the chairman of governors at the London School of Economics, had similar concerns when William Beveridge was appointed director of the London School of Economics in 1919. The latter created the Department of Social Biology in the London School of Economics in order to bring about the "application of biology to human society." The syllabus that sketched the disciplinary boundaries of the proposed department was outlined by Beveridge as: "Instinct in Man, Inherited and Acquired Characteristics, Quantity and Quality of Populations and Racial and Economic Tests of Fitness" (Dahrendorf 1995: 252). In making "quantity and quality of populations" central to his proposed new academic discipline, Beveridge was pointing out a troubling matter of public concern during those days, the problem of differential fertility.

Sidney Webb was attuned with Beveridge's conceptualization and adopted it in his socialist ideas. Speaking for the Fabian Society, Webb had told the National Birth Rate Commission in 1917 that rates and taxes fell most heavily upon the classes who should have most children but presented no impediment to "the thriftless and irresponsible, the reckless and short-sighted of all grades," with the result that "the community now breeds fastest from its least desirable stocks" (Soloway 1990: 155). Indeed, Webb had shared the eugenists' concern for possible degeneration of the racial stock resulting from the differential birth rate. The interesting thing is that the proliferation of the unfit was related especially to Catholics, Jews, and immigrants, who "bred freely." Sidney Webb in 1907 had argued not for flat rate benefits, but for the "endowment of motherhood," since the alternative was "this country gradually falling to the Irish and the Jews" (Semmel 1958).

According to H.G. Wells, "We cannot go on giving you health, freedom, enlargement, limitless wealth, if all our gifts to you are to be swamped by an indiscriminate torrent of progeny" (Wells 2004).

The way to alter this condition was, according to H.G. Wells, to provide free medical care for the "childbearing women of the appropriated classes." Social health, in the view of Fabian socialists, meant the encouragement and improvement of the healthy parts and productive elements of society, and contemporaneous removal of the sick and parasitic elements. Without these measures, which he outlined in *Socialism and the Family*, Britain would be directed towards race deterioration (Wells 1906). As we can see, the concept of ethnic exclusion was intertwined with the idea of non-productivity. In other words, the Catholics were the "non-productive." In 1902, H.G. Wells published a work titled: *Anticipations of the Reaction of Mechanical and Scientific Progress upon Human Life and Thought*. The last chapter in this book is entitled "The Faith, Morals and Public Policy of the New Republic." It is clear that this faith is not Christianity, or perhaps any other transcendentalist faith. Wells speaks of a new class of humans, which he calls "the dominant men of the new time." He writes:

> The ethical system of these men of the New Republic, the ethical system which will dominate the world state, will be shaped primarily to favor the procreation of what is fine and efficient and beautiful in humanity—beautiful and strong bodies, clear and powerful minds, and a growing body of knowledge—and to check the procreation of base and servile types, of fear-driven and cowardly souls. To do the latter is to do the former; the two things are inseparable.
>
> (Wells 1902: 322–323)

Wells glorifies a "class that has naturally segregated," who will have the "passion … to get the maximum result from certain possibilities." The Fabians were the key innovators of the use of eugenics for social engineering—to them, eugenics was a scientific instrument for reducing need. They employed positive eugenic policies, striving to improve the health of the population by increasing the rate of reproduction of the fittest with free medical care for the "childbearing women of the appropriate classes." With the same zeal they defended negative eugenics in order to reduce social margins.

Not surprisingly, Fabians were critical of the laws that lay at the basis of Britain's welfare state such as the 1834 Poor Laws. They emphasized the fact that these laws were regressive rather than progressive because they encourage childbirth of the inhabitants of urban slums.

The Fabians were not the only ones who considered the Poor Laws outdated; that is the reason that in 1905 the British government established a Royal Commission to look into "the working of the laws relating to the relief of poor persons in the United Kingdom." Beatrice Webb was asked to serve as a member of the commission, and disagreed with most of the members on the Royal Commission. With Sidney Webb she published a Minority Report that called for the end of the Poor Law, the establishment and coordination of employment bureaus throughout Britain to make efficient use of the nation's labor resources, and improving services such as education and health. The Majority Report, which represented the views of Helen Bosanquet, was based in the traditional individualist view that pauperism is a moral condition. Social policy, according to that view, should be directed towards the necessities of the individual and to foster a spirit of independence.

The Minority Report drafted by Sidney and Beatrice Webb shared the Majority view that increasing poverty and destitution were rooted in deviant moral values. Unlike the Majority, however, the Webbs considered "bad" moral character as a consequence rather than a cause of the wider public issue that concerns social disorganization. People became poor, according to Beatrice Webb, from a variety of causes—from old age, from ill health, from mental disease, from the loss of the breadwinner, from wages that were too low to support the family (Cole 1946: 110). It must be said that Webb's idea could have broken down a binary distinction between a tax-paying "us" and a welfare-claiming "them." How is this connected with eugenics?

As maintained by Sidney Webb:

> Poor Law, which was costing the United Kingdom nearly twenty millions a year, was almost entirely anti-eugenic in its tendencies. In its provision for feeble-minded maternity, and in the opportunities which the general mixed workhouse afforded for acquaintanceship between the sexes, the "Poor Law" had undesirable results from the eugenic point of view,... laissez faire, from the eugenic point of view, was the worst of all policies, because it surrendered the idea of intelligent, purposeful selection. The first consequence of withdrawing public provision would be an outburst of private charity of the most sentimental order, which would be beyond the power of any Government to suppress.
>
> (Webb 1909)

It thus seems to be clear that socialism Fabian style could not stand for Poor Laws and should be eugenic. As fully stressed by Edward Pease, the Minority Report of the Poor Law Commission—although it never mentioned socialism—was a notable and wholly original addition to socialist theory, entirely of Fabian origin. Hitherto all socialist writings on the organization of society, whether contemporary or Utopian, had visualized a world composed exclusively of healthy, sane, and effective citizens, mostly adults (Pease 1911: 1909–1915).

In a similar vein, Karl Pearson, a Fabian with Marxist economic leanings, claimed that the most fertile sector of the population produced 50 percent of the next generation (Soloway 1990: 168). The much delayed publication, in 1923, of the results of the 1911 Fertility of Marriage Census showed that, despite increasing resort to birth control by successive marriage cohorts since 1870, there was a persisting social class differential in birth rates. The problem was now officially recognized, as people used to say in those days, that "the rich get richer and the poor get—children."[25]

With the same zeal that they attacked *Poor Laws*, the Fabians defended the law of "minimum salary" for eugenic reasons. A minimum wage was seen to operate eugenically through two channels: by deterring prospective immigrants and also by removing from employment the "unemployable," who, thus identified, could be, for example, segregated in rural communities or sterilized (Henderson 1900).

It is not strange, thus, that the Fabians were democrats who believed that only a productionist society could be a social welfare society. The central characteristics of the future society were defined by Beatrice and Sidney Webb in their *History of Trade Unionism* (1894) and *Industrial Democracy* (1897). In *Industrial Democracy* they wrote quite clearly that "with regard to certain sections of the population [the 'unemployable'], this unemployment is not a mark of social disease, but actually of social health" (Webb and Webb [1897]1920: 785). In short, to Sydney Webb the most ruinous to the community is to allow "them" to unrestrainedly compete as wage earners (Webb 1912).

This view competed with neoclassical critics such as Alfred Marshall (1897), for example, who believed that job loss as a social cost of minimum wages is completely wrong. In other words, we can see here two versions of economic productionism. A free market, devoid of minimum wage, is open to immigrants who help develop the economy. As an alternative, the Fabians promoted minimum salary, which prevents immigrants from competing with nationals. These "nationals," however, should be induced to reproduce themselves if society strives to remain productive. With the same zeal, the non-productive elements should be discouraged from reproducing themselves.

Although the idea that Fabian socialism led to the promotion of policies of sterilization is open to debate; more than one testimony leads to acceptance of the version promoted by H.G. Wells after hearing Galton's exposition of "eugenic religion" before the Sociology Society in 1905, that failures should be sterilized.[26]

In sum, the merging of Fabian and eugenic views had been represented clearly by Karl Pearson, who combined a socialist view of the crisis of capitalism with a disdain for class struggle and with a nationalistic veneration of the State. Under a dictatorial socialist State, Pearson argued, struggle between classes would be replaced by struggles of the unified nation with weaker peoples. Is it not, he asked, "a cause for human satisfaction," that a "capable and stalwart race of white men should replace a dark-skinned tribe which can neither utilize its land nor contribute its quota to the common stock of human knowledge?" (Pearson 1892: 300).

This idea was addressed by Havelock Ellis (1859–1939), who in his book *The Task of Social Hygiene* (1912) argued that in order to progress a socialist society must eliminate its non-productive elements. In a similar vein, Eden Paul, the well-known socialist physician, claimed that unless the socialist is a eugenicist as well, the socialist state will speedily perish from racial degradation.[27] In more practical terms, they distinguished between a prudent organized working class and the lower classes, which were prone to reproduce themselves with such fecundity that they would outnumber the productive classes. This meant that they were ambivalent towards the poor. In some cases they were portrayed with sympathy and in others with contempt. They were in favor of eugenics while simultaneously promoting measures for improved health, education, and welfare (Kramnick and Sheerman 1993: 37).

The final question is: What was the Fabians' legacy? From a practical political sense they have not succeeded in passing eugenic laws. That might lead observers to dismiss the British eugenicists' importance. But legislation is not the only—or perhaps even the best—measure of success. Whole sections of British society were influenced by the Fabians in their conviction that talent and character were inborn and fixed. As Diane Paul argues, while citing Edgar Schuster and Ethel Elderton of the Galton Laboratory, at the time of the first publication of Mr. Galton's *Hereditary Genius* in 1869, the belief in the hereditary nature of inborn natural ability was held by very few. However, after the publishing of the book there was virtually no educated person that would dispute that claim. This assumption had consequences far beyond programs of eugenic selection. Paul also stresses that plays like Shaw's *Man and Superman* (1903) and novels like Wells' *A Modern Utopia* (1905) did more than academic studies to popularize the concept of selective breeding (Paul 1995: 75–76).

Interestingly, the most important thing is that the Webbs' research led to the conclusion that a model of industrial democracy could be achieved through democratic means.

However, this did not prevent the Fabians from flirting with the Soviet Union on the one hand and defending colonialist and nationalist policies on the other. Indeed they share a concept of imperialism rooted in "universal values" rather than in pure egoistic nationalism. Pease, for example, sustained the Fabians' support of the expansion of the Empire only insofar as it was compatible with the expansion of that higher social organization which this Society was founded to promote (Pease 1911: chapter 7). Probably for the same reason some Fabians, despite their defense of democracy, were impressed by the Soviet Union. Indeed, the organizational brand of Soviet state socialism lured the Webbs, who visited the Soviet Union in 1932. Although unhappy with the lack of political freedom in the country, they were impressed with the rapid improvement in the health and educational services and the reforms that addressed the need to secure economic and political equality for women. They considered the Soviet Union a new civilization and forecasted that the social and economic system of planned production would spread all over the world.

In other words the Soviet Union is portrayed as a productionist rather than a humanitarian socialist society. At the end of the day, the ideas of productionism as a key to undermine degeneration also fit the proletariat nationalism and Marxist scientificism of the Soviet Union.

Roland Daniels (1819–1855), influenced by Marx's *Poverty of Philosophy*, claimed that the task of science was to uncover the revolutionary elements in human misery that could act as the basis for defeating conditions of impoverishment. After all, a proletarian science aimed at liberating people from their bourgeois deficiencies and enhancing the creation of inheritable proletarian virtues. Science played a crucial role in promoting social progress. Progress, however, was not inevitable because under certain conditions the "unfit" might prosper, posing a challenge to any further evolutionary development.

These concepts were also popularized by Marxist revolutionaries. Lenin himself claimed that colonialism had given the advanced industrial countries a tremendous competitive advantage over the less developed, non-colonialist capitalist states.[28] Russia, Lenin concluded, was the weakest link in the capitalist chain. The phenomenon of national degeneration is best understood within a complex of beliefs that arose from a constellation of anxieties: the agriculture depression, the increase in proletarian discontent, and fears about foreign competition. Each of these concerns was visible in Lenin's ideological rhetoric of Russia degeneration: "One step forward, two steps back.... It happens in the lives of individuals, and it happens in the history of nations and in the development of parties" (Lenin 1960, vol. I: 470).

For Lenin, Russia's salvation depended on the resurrection of its working-class base and so recovering the nation's identity: the nation might know itself again through the deployment of the proletarian revolution. The prevailing sense was that the nation's incompetence was a symptom of pervasive degeneracy. The class struggle mirrored the efficiency of the nation:

> Only the revolutionary works. If they are supported by the poor peasants, they are capable of smashing the resistance of the capitalists and leading the people to gain the land without compensation.
>
> (Lenin 1960, vol. II: 231–232)

A strong communist society, thus is represented by the leading role of a healthy proletariat, followed by the peasants. In sum, in contrast to Nazi racist eugenics, which gives priority to racism rather than productionist goals, and slightly different from a Marxist concept of eugenics in the service of a proletarian revolution, here we have a concept of eugenics in the service of the welfare state.

Furthermore, it appears after analyzing the Fabians' rationale, that the ideological links of eugenics with democratic socialism is stronger than with Nazism, fascism, and communism.

By highlighting and exploring the construction of degeneration as a core component of ideology, each ideology found its balance differently. Hitler's national socialism, by stressing race and biological heredity as atomizing features, showed a tendency toward a dominant emphasis on peripheral concepts of ideology rather than on degeneration in its universal sense (Schiller 1921). Thus, the idea of *Rassen* hygiene (racial hygiene) and eugenics, as used in the German context, incorporated the concepts of genetic inferiority and superiority independent of anthropological race, and had a racist overtone of maintaining the purity of the "Aryan race" within the Nazi ideological discussion of public health (hygiene).

Conversely, Mussolini's fascism, Leninism, and Stalinism also gave priority to integrated peripheral concepts of youth, class, nation, and imperialism into a goal-directed and coherent ideological core rather than degeneration. Thus, the general trend of transformation from racial-biological concepts to social-environmental ones maintained the unity and relative stability of degeneration as an ideological core. Such a trend paved the way to productivist conceptualism of degeneration. By social-national hygiene we are to understand what was formally known as social reform, which concerns the welfare of human beings living in societies. Social-national hygiene was defined in terms of social and cultural productivity rather than in racial differences. Within this context, eugenics became at the service of social democratic ideology rather than a coherent ideology in itself, and consequently survived the transition from racial hygiene to national hygiene politics. This does not mean that eugenics did not fit liberal productionists or racists in the United States, for example. As Hansen and King (2001) remarked, eugenic ideas became politics depending on their use by political elites. Despite the fact that eugenic ideas were developed by the Fabians, in

Britain they did not hold. In the United States, in contrast, eugenic ideas were implemented by racist politicians in a particular time conjuncture.

The central point in our study, however, is that while a liberal concept of productionism can live without eugenics, it seems, as we shall later see, that when a social democratic productionism concept meets the nation, eugenic thinking is the only mechanism for national survival. The meeting point between socialist productionism (which could be democratic or not) and eugenics, thus, is ideologically and politically compelling. A welfare social democracy in people's-home countries, namely nation states, needs eugenics as a means of survival.

Summing up

This chapter focused on two contrasting ways of approaching eugenics: science and ideology. We have traced the applicability of eugenics to paradigmatic changes in the fields of anthropology, human genetics, and ideology with which eugenicists attempted to gain paradigmatic dominance.

Drawing on disciplinary history, it is possible to conclude that eugenics could no longer exist as both a paradigmatic scientific discipline and a coherent and stable ideology, but rather as an instrumental ideology of degeneration that could easily support dominant ideologies such as nationalism and socialism in the post-World War II era.

Within the scientific disciplinary framework established by anthropology and human genetics disciplines, eugenics was unable to fit the paradigmatic transitions. Early eugenics invoked anthropology to establish a link between race and hereditary biology. However, the reorientation in anthropological studies of human races signals the crystallization of a paradigmatic consensus that the concept of race had no biological validity and that the many intermediate grades of variation between races made the distinction between them arbitrary.

This paradigmatic shift positioned eugenics on a cracked ground, which led to eugenics being regarded as a distorted judgment about how inferior races ought to be treated within society as an accepted paradigm. In the case of human genetics, paradigmatic changes showed a break with the Mendelian explanation of human intelligence and morality. These variables were far too complex to be understood in simple Mendelian terms. Such a paradigmatic shift proved to discredit eugenics as a legitimate scientific endeavor to explain heredity within human genetics research. In fact, when the modern synthesis established itself as a permanent feature of the genetics landscape, eugenics could no longer provide a conceptual territory within the biological sciences in which the discipline could operate. Finally, during the 1930s, there was considerable agreement among geneticists about what constituted the paradigmatic core of a genetic discipline, leaving eugenics outside the discipline.

Within the ideological discipline, eugenic concepts could not support claims for a well-constructed ideology. Degeneration as a core component of ideology took a peculiar form in eugenics in that it insisted on an essential identity of biological race and degeneration. Eugenics even went so far as to anticipate the assimilation of degeneration into a biological discourse of race in order to construct its ideology. However, the leading political ideologists of the 1930s had shifted from claims about the presumed link between degeneration and biology. In Mussolini's fascism, Leninism, and Stalinism's social-environmental concepts such as youth, class, nation, and imperialism were lifted out of their defining context and contributed to revitalizing the constant fear of degeneration, which could remain intact to fit changing political realities.[29] Thus, by placing race and hereditary features within its ideological core position, eugenics as an ideology entered the phase of an inconsistent ideological structure (Schiller 1921).

However, this chapter shows that in contrast to the dismissal of eugenics from the scientific disciplines, eugenics was proved to survive in the ideological discipline. On the surface, it seems that eugenics' hold on racial "degeneration" did not become a dominant ideology such as fascism and Leninism. However, its unstable ideological core of degeneration did help eugenics to survive the transformation from racist ideologies to a social-environmental one.

In this chapter thus it is argued that eugenics' ideological flexibility found a home in the emerging democratic socialism. As the idea of degeneration became important in democratic socialism, eugenics' main tenets were embraced by socialist reformers. This is a link that is seldom discussed. In its initial stages, democratic socialism was constructed to respond not only to capitalist economy but also to the moral shortcomings of both the capitalist system and Marxist-Leninism proletariat dictatorship, by propounding universal moral values.

At first sight, the idea of degeneration appears alien to democratic socialism, whose basic ideal should be the advancing of egalitarian social welfare. However, as we have analyzed, a new trend in democratic socialism, a productivist socialism as projected by the British Fabians, appropriated the issues of degeneration and eugenics. By tapping into the intellectual roots of eugenics and explaining its connections with productivist socialism, we underscored in this chapter how eugenics' role has been shifted from racial hygiene to social-national hygiene. Eugenics was employed in social hygiene policies by productivist socialists, who associate the concept of productivity with health. The productive sectors of society were the healthiest ones and, of no surprise, these healthy productive sectors of society were the "dominant" ethnic group.

There is no question that this setting could be associated with racism. However, as we have noted, although the racist connotation of productivist eugenics is unavoidable, the basic aim of this productivist approach was to promote welfare for all, including the weakest sectors of society. In more senses than one, the productivist welfare socialism set the basis for an exclusionary form of incorporation. The condition of incorporation was productivity, and under the same parameters, the feeble or those in the social margins would be forbidden to reproduce and grow.

Thus deprived of any scientific basis, eugenics continued to flourish entangled in a productivist social democracy rather than Nazi racist systems.

The aim of protecting the national stock, a national culture, without falling into overt racism, added to the aim of promoting a welfare industrial productivist democracy, leads to eugenic policies and to genetics in present times. Such an ideational account of eugenics as put forward in this chapter will provide a framework for the following historical and current accounts of eugenics across population policies in welfare nations. Thus, in more senses than one, the eugenic productivist approach set the basis for an exclusionist or conditioned way of inclusion.

Saying that, we can proceed by showing how labor and social democratic movements in three of the case studies we deal with—Sweden, Israel, and Demark—developed eugenic institutions in the 1930s and 1940s and how this is connected with genetic policies developed in present days by a wide variety of welfare democracies. Of no question, social democratic ideas were the underpinning of well-structured industrial democracies, and as Sherry Berman notes, it is Swedish social democracy that represents social democracy in its pure character. During the interwar years, social democrats displayed a new strategy based on state control of the market and communitarian appeals. However, it was only in Sweden that the main party of the left embraced this new course wholeheartedly, and it was thus only in Sweden that socialists were able to outmaneuver the radical right and cement a stable majority coalition, setting the foundation for a period of political hegemony unprecedented in democratic societies (Berman 2006: 152).

While German social democracy plunged into doctrinaire debates, the Swedish social democrats succeeded in promoting a common formula blending democratic nationalism expressed in the idea of "peoples home" and a technocratic mind. We might add that Sweden set a type of democratic productionist socialism that set a welfare system for the strong. That would constitute a middle road between a universal welfare that makes the needy permanently dependent on the state, and fascism's anti-democratic nationalist productionism.

As can be grasped, the Swedish idea of a welfare industrial democracy was embedded in eugenics ideology, and became a landmark especially for Nordic welfare societies. A similar case would be that of Denmark and Israel during the Labor movement's predominant role during the nation-building era. It synthesized nationalism with a modernist technocratic ideology which basically should shape a new "Jewish character."

We add the example of Finland, despite the fact that the welfare state and eugenics were developed by a national coalition in which social democrats belonged but were not the most prominent agent. Finland, however, could be defined as a Lutheran social democracy, implying with that a "conceptual national unity" on the communitarian values of welfare productivism. Similar to Israel, productionist eugenics develops first as a key for the survival of an ethnic group before it becomes nationalized. As in Israel, the development of the welfare state was totally entangled with the raising and reproduction of the healthy and productive associated with a particular ethnic group. Differently from Israel, however, eugenic productionism in Finland evolved into a whole Finnish national ideology. In Israel it came to the service of the Jewish ethnic group.

In all of these cases but one, social democracy as a first promoter of the welfare state was totally embedded in eugenic productionism.

2 From “race hygiene” to “national-productivist hygiene”

Introduction—an exclusionist path to inclusion

This chapter examines a little-known episode from the history of the labor movements in Sweden, Demark, and preindependence Israel, as well as from the development of the Finnish welfare state. These examples offer a fruitful way to expand our understanding of the shift of eugenic efforts from racializing to productivizing collectives in people’s-home nations. In this chapter, we explore the way in which eugenics shifted its emphasis from what was referred to as “race hygiene” to “national productivist (social) hygiene” during debates in the late nineteenth and early twentieth centuries, and analyze how that shift was applied in Sweden, Denmark, Finland, and Israel, all welfare nation states in which social democracy or the labor movement (except in Finland) had a leading role in displaying eugenic measures designed to craft a welfare society for the strongest and healthiest parts of society.

We pick these four countries not only because they were welfare societies, nation states, or “people’s home” countries, whose socio-economic development relied upon national productivist eugenics. They are also important because the four expose in current times the most permissive examples of ART; Sweden, Israel, and Finland coming first and Denmark following suit.[1]

Why is this so and what are the relations of these political examples with their eugenic pasts?

We argue that there is a continuing line or “policy convergence” between the national productivist eugenic regime of the past, which was synthesized in the enacting and legitimization of eugenic legislation, and current permissive programs of ART.

A large variety of theories explain the origins and sustainability of the Scandinavian welfare state. According to Hugh Heclo and Bo Rothstein, for example, the early formation of corporatist institutions of compromise in Scandinavia explains the development of the welfare state (Heclo 1974: 315; Heidenheimer *et al.* 1990; Mahoney 2000; Rothstein 1988). Slightly different from the last claim, we stress the “productivist” ideology behind the creation of the mentioned welfare societies, and suggest that the noted productivist society was reinforced by eugenic measures promoted in most cases (although not only) by labor or social democratic movements.

At first glance there is an inherent temptation to relate the introduction of eugenic sterilization to the Scandinavian work ethic and Lutheran connection. While we do not dismiss this claim, we stress the role of ideas, especially the productionist ideological background behind eugenic practices. It is true though, that the ideological framework of productivist eugenics is akin to nation states or to states that strive to protect a hegemonic national culture.

We suggest that these welfare-productivist-eugenic settings have become part of a societal technocratic national culture that is projected in time at different political junctures.

Differently from racist hatred, the welfarist productivist idea has set an exclusionist or conditioned path of inclusion that tends to reduce the social margins. Inclusion implies productivity. In present times we see a reemergence of an updated face of the welfarist-productivist-eugenic ideology under new socio-economic constraints.

The question for these countries is how to preserve a welfare people's home under a globalized open economy and under huge immigration waves.

As we have stressed in the first chapter, unrestricted immigration might be economically advantageous to any country, but in the countries we examined there is a desire to preserve a societal-national identity, diminishing the impact of immigration while at the same time being economically successful in a globalized economy.[2]

We suggest, thus, that in order to fulfill these inherent desires to survive as productivist welfare nation states, the politics of eugenics guised under a different label becomes enticing. To be precise, the continuity path of eugenics into ART should by no means be identified with eugenics because of the bad reputation of eugenics in the past. However, as we shall try to prove, the eugenic mentality is embedded in the welfare productionist mentality of the countries analyzed.

We suggest that in the four examined countries: (1) the eugenic trend went through racial hygiene towards a national productionist ideology; (2) social democratic and labor parties' policies had an important impact in the early stages of this process; (3) finally, despite the fact that we are dealing with liberal open societies, the examined countries converge in their population policies.

They converge in ART regulation and coverage, tough immigration policies (in Israel immigrants are generally Jews; work-seekers without the necessary visas are not accepted and are expelled), and family planning.

Indeed, at least in the countries studied, we found a continuing line or "policy convergence" between the national productivist eugenic regime of the past, which was synthesized in the enacting and legitimization of eugenic legislation, and current permissive programs of ART.

Despite important differences, Sweden, Israel, Denmark, and Finland all defined themselves in different ways as nation states or "people's home" countries that employed eugenics under the guise of "productive welfarism."

The connection between eugenics, productivism, and welfare is thus framed around several forms of concern about population in the four countries: sterilization, restricted immigration (of non-Jews in the case of Israel), and reproduction.

In Denmark and Sweden the people's home coincided with the productive people, which sustained the welfare state. In Israel, the productivist identity coincided with one ethnic group, Ashkenazi Jews, at the early stages of nation building and afterwards it included Sephardic Jews, although it never included Arabs. In Finland, after the Civil War, the productivist identity evolved into a whole national culture including Swedish-speaking Finns and the Finnish majority, bourgeoisie conservative parties, and Red social democrats. Different from Israel, however, in Finland what began as a conflict between these two ethnic groups and between Whites against Reds evolved into a whole Finnish national identity in which the eugenic trend became an integral part of its development.

In the previous chapter we presented the idea of "racial hygiene," which maintained that biological differences among races would inevitably lead to national degeneration. This thesis was logically connected to Nazi racial politics and largely with fascism during the 1930s and 1940s. However, as we suggested, and this is our main point, social democrats also made use of eugenic politics in order to enhance their national stock.

How do we make sense of this?

Indeed, common sense observers may assume that social democratic movements were supposed to present not only an economic productionist response to capitalism, but also a moral and democratic alternative.

In this sense, social democratic movements endorsing eugenics sounds strange at first glance. Indeed, while social democrats and labor movements had always been associated with universal welfare and distribution, few heeded the productivist aspects of social democratic movements and its ensuing development into eugenic policies.

This is what this chapter is about.

By remarking on the shift from racial to "national-productivist hygiene" and analyzing how it was intellectually and politically processed, we shall see in political praxis how the very idea of welfare productionism, elaborated by social democratic movements, set the ground for an intermediate between racism and liberalism, basically tailored for people's-home countries. Indeed, the very idea of a "home of the people" country is representative of a welfare society built as a unique experiment in social engineering.

In three of the four countries examined, their labor movements were totally engaged in engineering a welfare society, which in order to survive had to invest in its productive people. In Finland, the Social Democratic Party's share of the popular vote was much less than its counterparts elsewhere in Scandinavia, perhaps partly as a legacy of the Civil War. However, its contribution to the establishment of Finnish welfarist and productionist national identity cannot be disregarded.[3]

Similar to Fabians, Swedish Social Democrats, Danish Social Democrats, Finnish of all political sectors, and Israel Laborists promoted eugenic ideas on the basis of technocratic, rather than racist, reasoning.

However, as we shall see, this technocratic trend in reality disguised one single goal: the preservation and enhancement of the national community, because it is the one that can preserve and lead the productive enterprise. Indeed, population control policies were utilized by the Swedish and Danish Social Democrats, Israel Laborites, and Finnish Conservatives and Social Democrats, as a technocratic tool to preserve the productivity and security of the nation rather than preserving the purity of race. This admixture of nationalism and productivism was reflected in negative and positive eugenic policies. Negative eugenic policies included voluntary and compulsory sterilization and stricter control of immigrants, while positive eugenic policies included encouragement of reproduction of the "right living" through child allowances and tax exemptions for families with children.

In all of these cases, technocratic and modernist eugenics rather than racism was behind the eugenic drive. In all of these productivist societies, large or small labor and social democratic movements paid lip service to universal welfare and universal rights, while the welfare of "productive" people was advanced at the expense of "feeble."

In our historical comparison we should start by explaining how productivist socialists generally associate the productive sectors of society with the "dominant" ethnic group, and in this sense they become associated with nationalism. Indeed, *we should stress* that when nationalism, or at least the myth of a national community, becomes the engine of productionist socialism, eugenics comes to the aid of the national community in its aim to reproduce itself.

Leaders in all these countries shared a common pattern of thought, blending universal values and the productionist part of socialism, while adding a deep preoccupation for national survival. The solution was to be a synthesis of the ethnic with the productionist, when the productionist formula permitted escape from an utterly ethnic definition. After all, the right ethnic group should grow healthy and expand even at the expense of the feeble who belong to the right group. Eugenic policies used in a seemingly democratic manner thus were meant to enhance a healthy society as a means of survival.

The evolution of Swedish social democracy as a technocratic and democratic movement totally determined to create a modern and efficient industrial democracy, is at the basis of its choice of eugenic policies as a tool to promote precisely those goals. In Israel, the creation of a modern Jew, a pioneer and a fighter, was symbolic of the new healthy society. The Labor movement was essential in crafting this model of healthy fighter-worker citizens. In a technical sense, there are very small differences between Denmark and Sweden in terms of the structure of state and society. Social Democrats used eugenics in order to craft a productionist program that was associated with the "Danmark for Folket" (Denmark for the People) program, which set a middle way strategy that aimed to mobilize all social classes and political parties around the social welfare project. Since we are talking about a homogenous society, eugenics came to the assistance of ethno-nationalism defined as a productive society.

In Finland, as we shall see, eugenics was first developed by the Swedish-speaking Finnish minority; however, with the development of a Finnish inclusive national identity, eugenics became a tool for preserving the nation. Breeding Finnish people in a more insightful way was a step further from the previous idea that education and scientific education serve the same productionist goals.

Finally the question again is whether we should clean racist exclusionist connections from these societies. There is no question that from a current liberal multicultural perspective, the national productionist version of exclusion can be defined as a racist one. However, from a historical perspective, a racist approach implies excluding and somehow exterminating the "others," while the national productionist version enhances the strong group within society for the benefit of all.

In other words, slightly different from a purely racist approach, the national productionist version of eugenics attempts, through enhancing the right stock, not only to protect it but to create benefit for all, including all ethnic groups. In more senses than one, the productivist concept disguised the idea of racist exclusionism into a more technically meritocratic concept.

The labor movements in Israel and Social Democrats in Sweden and Denmark did not exhort racial hatred. However, there is no question that the Israeli Ashkenazi regarded Oriental Jews as inferior and needing to be assimilated. Similarly, the Swedish regarded the Sami minority as inferior.

That might be the reason why in the 1960s, as a way to mark a change with the past, Sweden adopted at least theoretically multicultural policies. However, in reality, both Sweden and Denmark remain people's-home countries, implying with this that the question of numbers of immigrants challenging the productivist collective identity does make a difference. In a supposedly multicultural Sweden or in an assimilationist Denmark, the idea is that foreigners should integrate themselves into the norms of a productivist mentality. In any case these are exclusionist policies, especially those directed against the feeble.

In present times, the feeble in three of the studied societies, Sweden, Denmark, and Finland, are represented by immigrants who are dependent on a non-productive "universal" welfare state. Since the idea is to diminish the number of dependents in the welfare system, the productivist nation needs more breeding of the productive national stock and less immigration. In Israel the question of demography comes as an imperative issue for the Jewish majority vis-à-vis the Arab minority and lately vis-à-vis the growing number of illegal immigrants coming from Sudan and Eritrea. In Sweden, Denmark, and Finland, while lip service is paid to diversity, as we will show, the attempt is to increase the breeding of the national/productivist people.

In more senses than one, the countries selected for this comparison are four of the most generous, inclusive social democratic welfare states: the four countries developed high levels of welfare inclusiveness for what they defined as the productive people, and in present times they set up a high level of ART programs for their "productive" people.

Sweden: from race hygiene to welfare and production

Swedish social democratic exclusionism

Before entering into the connection between Swedish social democracy and eugenics, we should stress that race biology dominated genetic science and eugenic rhetoric in Sweden before World War I (Broberg and Tydén 1996).[4] The ideology of eugenics in Sweden was sustained by a long intellectual heritage of physical anthropology. In 1882, a growing number of studies of the Swedish population were performed on behalf of the Swedish Society for Anthropology and Geography, conducted in light of the marked changes in population due to migration. In 1909, the Swedish Society for Racial Hygiene was established in Stockholm, and the Mendel Society, the first Swedish genetics association, was established in 1910. Before World War I prominent physicians such as Herman Lundborg proposed viewing eugenics as a reasonable answer to the problematic influx of foreigners. The most representative example of this trend could be synthesized in the proposal advanced by F. Lennmalm, a professor at the Royal Caroline Institute of Medicine, to establish a Nobel Institute for Racial Biology.

Although the proposal was not enacted, a bill introduced in both chambers of the parliament in 1921 promoted the creation of a state institute. Lennmalm's proposal was accompanied by wide support from a broad spectrum of parties from the conservative right to social democratic left.

In any case racial hygiene, propagating the racial quality of the Swedish population, was advocated by representatives of all parties in parliament and especially by Alfred Petrén, a Social Democrat member of parliament and head inspector of all mental care institutions in Sweden. Alfred Petrén strove to regulate the cases of sterilization that were not based on strictly medical grounds. He was motivated specifically by racial hygienic reasons and he pointed out three groups who were forbidden to marry according to Swedish marital law: the mentally deficient, the mentally ill, and the epileptic.[5] In this sense we can claim that he was a forerunner in demanding a shift of the sterilization issue from genetics to the social arena.

An act that may portray and synthesize Alfred Petrén's attitude towards the issue of sterilization can be seen when in 1933 he issued a motion to parliament by which he criticized a law on sterilization presented by the government, because the latter demanded that an individual consent to be sterilized (*rattskapabla*).

He held that the bill, in its original formulation, failed to articulate the interest of the state, which, as he declared, was to prevent mentally deficient persons from having children. The need for consent would unduly reduce the number of operations that could be performed.

The question that must be asked thus is whether Alfred Petrén's example is or should be confined to the fringes of Swedish social democratic praxis. The answer is negative. Alfred Petrén's initiatives signified a gradual shift of reasoning about and managing sterilization of the mentally deficient from racial to social-productive categorization, which fit the social democratic goals of defining the boundaries of a Swedish people's home. The definition of "productive quality" and of the "socially marginal," who had to be eliminated because of their high social cost, became important during the 1930s. Such a definition was used by Social Democrats to locate their concern for the genetic well-being of the "collective" rather than of the individual. This "collectivist" label for eugenics doctrine or policies treated the individuals who constitute the "socially marginal" group on the basis of the contribution that they might make toward the well-being of the collective.

This raises the question of how Swedish social democracy reached this point. In the debate on those matters in Sweden in 1977, Maciej Zaremba claimed that

> … purging the "inferior" was mainly a social democratic project. Neither was it a temporary aberration or a concession to the spirit of the times. In Sweden sterilization surgery formed an integral part of the very base of the construction of the welfare state.
>
> (Ekerwald 2001)

If that is so, the question is how that situation evolved.

As the twentieth century approached, Swedish social democracy promoted a middle way between German social democracy, which gave priority to theory over practice, and the English absence of theory (Berman 1998). They believed that politics is a problem-solving matter rather than ideologically conceived. However, despite absence of a clear ideology, Swedish social democracy stood in favor of two important principles—political democracy based in the right and obligation to universal suffrage, and social democracy, which relied on the central role of the unions. It can be argued thus that in Sweden, socialism and the labor unions are inextricably bound, and that differs from other countries where the ideology of one can be studied separately from the specifics of the other. Probably more Lasalleans than Marxists, they believed in the option of bringing socialism through a democratic conquest of the state.[6]

The role of the state on one hand, and the unions on the other, allowed for the classical formula of productivist socialism. To that, we should add the idea of people's home and of the successful association of workers with farmers. As Axel Danielsson argued, Swedish Social Democrats needed to move from ideological support for the working class to the people, the peasantry, and "low middle classes" (Danielsson 1972: 28–29). The coalition of the modern proletariat and the more traditionalist agrarian elements in a democratic coalition was quite promising and totally fit the idea of a Swedish people's home, a metaphor for the "supportive social democratic state" to full employment and full equality. Welfare nationalism was thus the idea of the great welfare coalition. In order to survive this, the welfare coalition had to define "productive quality," and the "socially marginal" became important. While such a definition did not need to imply a racial grouping, Social Democrats initiated the definition of productive people as the only "authentic" Swedes. Non-productive Swedes included not only foreigners but also categories of "feeble" Swedes, such as alcoholics. Thus, while Swedish social democracy did not promote racist hate, it certainly defined a "healthy" society as coterminous with the national stock.

Swedish social democratic productionism

The very idea of creating a healthy socialist society was in the mind of very prominent intellectuals. As with the Fabians in Great Britain, in Sweden political and social theorists set the ideological basis for eugenic policies and societal engineering.

The Swedish version of intellectual justification for social democratic use of eugenics could be found in the analysis of the population question as advanced by Alva and Gunnar Myrdal. Both academics, he a Social Democratic member of parliament, they authored a book that generated heated debate when it was published in 1934. Indeed, if any book could be seen as an authentic manifesto of the Swedish welfare state it is *The Crisis in the Population Question* by Alva and Gunnar Myrdal. It was published in Swedish and never translated to English. Observers, however, suggested that the book *Nation and Family*, written by Alva Myrdal, was supposed to present the same argumentation. In reality, the crisis of population defined the problem and the later *Nation and Family* set some solutions. In *The Crisis of the Population Question*, the Myrdals analyze the tendency of depopulation in Western societies in three chapters: "Malthusianism and neo-Malthusianism," "General Population Theories," and "Current Tendencies in Swedish Population Development." Although critical in general of neo-Malthusian theories for not capturing the question of the low birth rate of the 1930s, the Myrdals agreed with some of the latest neo-Malthusian proposals: the abolition of the law against contraceptives and the social regulation of abortions. In this connection they also wrote about unmarried mothers who are in need of support.

> There is something morally sick in a society that allows the social and economic boycott of unmarried mothers ... Is our country so poor and overpopulated that the birth of a child in so many cases would really mean economic destitution?
>
> (Myrdal and Myrdal 1935: 60–61)

Alva and Gunnar Myrdal oppose race and class hereditary theories; however, they could hardly accept that immigrants would be included in the nation. The Myrdals claimed that decreasing breeding figures had to be actively combated with positive welfare measures or the cultural and social integrity of the Swedish people would be threatened. The fourth chapter, "The Opinion Situation on the Population Question for the Near Future," argues against solutions to depopulation such as increased immigration. Not until social policy has made it less costly and less cumbersome for women to raise children can the depopulation tendency be overcome. The authors warned that a growing influx of foreign elements and the resulting lowered social standards would be the immediate result, and the working classes would be the first to be affected. This was the so-called "population crisis." According to the Myrdals, management of the population crisis should follow these assumptions:

> We are not interested in national expansion. On the other hand, we have said that a rapidly shrinking population, inheriting a rather rich and roomy country, must attract immigration. Immigration to an old country with a well-organized labor market and a rather highly developed structure of social welfare is something which probably does not occur without international

> friction.... This fear has, it is true ... been mingled with a mild sort of nationalism. After all, we in Sweden are all striving to build up a social and cultural structure of our own, better than the one we inherited.
>
> (Myrdal 1938)

Social Democrats began to promulgate eugenic views during Swedish parliamentary debates on sterilization in 1934. The striking thing is that the question of sterilization was also debated. Sterilization was suggested by Social Democrats as a solution to specific social ills. As a result of these debates and this ideology, a sterilization bill was enacted in 1934 authorizing sterilization of the "mentally ill." In 1941 it was extended to allow sterilization to combat "antisocial behaviour."

In 1941, the reforms promoted were more far-reaching than eugenic argumentation would allow, and were driven in part by legislators whose frustration resulted from the limited extent to which sterilization was performed under the existing legislation. The primary reason for expanding the law was to regulate the use of sterilization for those considered legally fit to give their consent to the medical procedure. The new law would permit voluntary sterilization of persons of "legal capacity." The proposed law added a social reason to the existing ones for sterilization, which was phrased, "... due to asocial way of life, is for the future obviously unsuitable to have custody of children." "Asocial" referred to such behaviors as vagabondry and alcoholism. Simultaneously, the administrative structure was changed, centralizing most of the decision-making to one board, which included two government political appointees, a clear indication of the socio-political agenda of the board.

From the humanitarian point of view, the central claim was that sterilization would protect children when one or both of the parents were categorized as "inferior." It was argued that these children would grow up in an unfavorable environment and not receive the care and upbringing necessary to develop into capable members of society.

In other words, social misbehavior was considered a question of genetic inferiority, assumed to be hereditary. Nils von Hofsten, the most important researcher of the Racebiological Institute, recommended that the National Board of Social Health and Welfare consider asocial behavior as part of heredity genetics. Some politicians opposed the idea that social behavior depended on genetics because they saw it as a straightjacket that would limit their legislative options for social engineering. For example, Oscar Olsson, a Social Democratic politician, played down the idea of biological inheritance and talked only of "social inheritance." He suggested that, as far as possible, the reproduction of asocial individuals who lead a socially destructive life should be prevented. Accordingly, the parliamentary debate in 1941 featured expanded social arguments that had partly "outgrown" the science of eugenics. In general terms, all the parliamentary debates on sterilization were characterized by a growing concern for the "deviant" or the "asocial," and the politicians legitimized their actions by promoting the general public interest. Critics even pointed out that the advance of eugenics led directly to sterilization policies.

This was, for example, a primary argument whenever the debate touched on the Tattare, a name assigned to transient groups such as the Sami population. The Sami were considered an aboriginal group with primitive manners, so that they needed coercion to be assimilated. According to Roger Kvist, from 1846 to 1971, the Swedish state "in practice created a system of institutionalized racism towards nomadic Sami" (Kvist 1994: 203). In Swedish folklore, the Tattare were supposed to be immoral and idle, with a dark and southern appearance. In the emerging Swedish welfare state the Tattare were an anomaly that belonged to the periphery of society. Individuals displaying social character traits generally associated with Tattare were assumed to be members of that "group." In the nineteenth century, in Sweden, itinerant travelers living on the margins of agrarian society were increasingly identified as a racial category of Tattares (Broberg and Tydén 1996: 124–125). However, in the early twentieth century, the Royal College of Physicians' classification of "intellectual"—as opposed to "moral"—feeblemindedness used the term "Tartar" to refer to what it otherwise called the "mongolian" or "kalmuck" type to which the question of itinerancy was not relevant. Broberg and Roll-Hansen (1996) argue that the notion of the alien Tattare as a biological reality and as a biological as well as a social threat was strengthened between the 1920s and 1940s.

Since their genetic taint was still unproven by more accepted scientific measures, the social legitimacy behind the Sterilization Act of 1941 was to end up with the social maladies attributed to the Tattare. An anthropometric study carried out by Gunnar Dahlberg with the collaboration of the State Institute for Race Biology in 1944, reached the conclusion that it was objectively impossible to separate Tattare from other Swedish citizens. In 1942, Dahlberg examined 66 male travelers recruited from asylums for the poor, prisons, and hospitals for alcoholics, compared them with the Swedish population, and found no significant differences. However, the general opinion was that this group of travelers was genetically and socially degenerate (Broberg and Tydén 1996). If we extrapolate these numbers, we might estimate that some 700 sterilizations were performed due to the individuals' ethnic heritage. In other words, if the Tattare were to be distinguished from the rest of Swedish society, a eugenic argument was useless. Thus, policymakers and administrators focused on social characteristics generally associated with marginality that came to be associated with "members" of the group (Lynöe 2007).

However, the case of the Tattare did not remain sound. Several voices were raised, calling for careful examination of the legal and moral problems of that policy. For example, parliamentary criticism was raised on the issue of the voluntary nature of the law. Georg Branting, a Social Democratic member of parliament, argued that forced sterilization was the real aim of the speakers in parliament. He suggested that there is a reactionary thought behind the regulation that persons who are considered unsuitable to have custody of children should be sterilized (Spektorowski and Mizrachi 2004: 350). Branting sought a different solution to the social problem. "As I see it—and I assumed that social democrats in general thought it to be evident—it is the duty of society to improve these miserable social conditions, not just for the parents but also for the children." However, in most cases, if there were objections, they did not pertain to the principle of sterilization as such but concerned mostly the due process of law in the implementation of the proposed bill. More widely supported than Branting, most critics of the existing law advocated methods of enforcement. Members of both the first and second chambers of parliament signed motions in favor of forced sterilization. A majority of these signatories were Social Democrats, and a large number of them were women.

In brief, the sterilization law enacted in 1941 extended eugenic considerations to include not only individuals suffering from mental retardation and defects of a hereditary nature, but also "antisocial" behavior as well. Assuming the prerequisites of eugenic or social sterilization were present, the operation could be performed without asking for the consent of the patients if they were considered unable to exercise their legal capacity.

By 1947 the number of sterilizations had grown to more than 2000 a year and the striking thing is that they remained at that level into the 1950s. Between 1935 and 1975, something like 62,888 sterilizations were performed in Sweden as the result of these laws.[7] The reason was that sterilization law and policy were not passive; Swedish authorities threatened those they wished to sterilize with extreme measures if faced with refusal, such as taking away children or denying abortions. Young people in reformatories had to agree to sterilization as a condition of their release or for a leave of absence.

The attempts of Swedish Social Democrats to employ negative eugenics did not end with sterilization policy and restricted immigration. Alva and Gunnar Myrdal also suggested employing a "positive population policy" followed by the eugenics rationale to improve the quality of human resources (Myrdal 1968). Based on economic and social reasons, the Myrdals suggested that it was seen to be necessary to "circumscribe the reproductive freedom of these lightly feeble-minded." The good of society, or "popular quality," as the Myrdals called it, preceded the good of the individual, and meant preserving its character as a people's home. The Myrdals wrote that "The direct task of prophylactic social policy is creating better human material." On the specific question of sterilization, Alva Myrdal added:

> In our day of highly accelerated social reforms the need for sterilization on social grounds gained momentum. Generous social reforms may facilitate home-making and childbearing more than before among the groups of less desirable as well as more desirable parents.
>
> (Myrdal 1968: 215)

In the 1950s and 1960s, the arguments that had sustained the old type of eugenics gradually faded from scientific debate in Sweden. However, that does not mean that sterilizations ended in the 1950s. When child allowances—monthly payments to families for each child, administered by the tax-funded national social insurance scheme—were introduced in the 1950s, the number of forced sterilizations of the "undesirable" part of the population doubled. Reacting to the defeat of national socialism, as well as to the postwar rise in the standard of living, which enabled greater individual freedom, Swedes reflected anew on the use of social eugenics. In general terms, the new mood was marked by a return to medical eugenics, and the question of sterilization was frequently connected to an altered view of the relation between state and individual (Spektorowski 2004; Spektorowski and Mizrachi 2004).

The national rationale thus became less prominent in the debate of population policies (the term "population policies" was discredited and was replaced by "family policies" in the 1950s), which pointed to a change from the general level to the level of the individual. Despite this change of mood, it was not until 1967 that the National Board of Health admitted that the policy of sterilizing learning-disabled people had been implemented because of an exaggerated belief in the importance of heredity, legitimized on social grounds.

It is obvious then that Swedish social democracy disappointed many observers who had hoped it would be a moral as well as an economic alternative to a free market economy. Those observers found it difficult to accept the Swedish implementation of sterilization policies and its theoretical backing as the basis for welfare reforms.[8]

In other words, Swedish social democracy ended up closer to national productionist socialism than a moral alternative to liberal capitalism. Eugenicists in Sweden, unlike those in Germany, tended to favor sterilization and segregation over euthanasia and more extreme "solutions" to national survival. In fact, the basis of negative eugenics in Sweden was to reduce the biological contribution of the most non-productive populations including non-Swedish, as well as those singled out as criminals, deviants, or mentally ill persons. The question is why there was no reaction by the Swedish people, especially after World War II. As we noted, the *folkhem*—as the concept is termed in Swedish—was to become the stable foundation of decades of peaceful labor relations, far-reaching social reforms, and unprecedented economic growth. The dark side of such reality was the harsh demand for conformity. People who did not correspond to the ideal of this new society were not welcome. Moreover, there was no court to investigate the constitutionality of new laws. Sweden has never had a revolution, and the state and its administration have enjoyed a trust that many foreign observers find astounding. There were no perceived reasons for the necessity of an elaborate system of checks and balances as in other countries. Thus, in a country with a relatively homogeneous population, with strong traditions of collectivism and consensus decisions, majority rule in parliament has been considered to be sufficient assurance that laws that are passed are on behalf of the nation's best interest.

Israel: welfare for the "new warriors"

The Zionist version of intellectual justification for the use of eugenics by Labor Zionists could be tracked back to Nachman Syrkin (or Nahman Syrkin, 1868–1924). Syrkin, who was one of the political theorists and founders of

Labor Zionism, provided an unusual perspective from which to analyze the project of social engineering that underlay the campaign to settle Jews on the land. Syrkin emphasized the need to view other natural categories that are rooted in diverse identities, such as racial, mental, and physical, as nature's first step towards the creation of divided groups of *Homo sapiens*. Following Syrkin's post-Marxist view of socialism, class struggle and confrontation do not produce tangible results; it provokes defeats and fails to solve immediate problems such as the "Jewish national question": "From prehistory past … the appearance of various species of Homo sapiens evolved with racial, mental, and physical characteristics deployed to subordinate particular groups of people" [my translation, L.I.S.] (Nachman 1929). Syrkin's use of racial articulation of the nation was keyed to two fundamental realities that confronted Zionist socialism: first, the inherent ambivalences in Zionist nationalism and, second, the nature of the socialist project.

Explicit and implicit views of race were of importance primarily in the German branch of Zionism. The Zionist adherence to a racial eugenics approach was increased during the first half of the twentieth century. The racial paradigm predominated Jewish race discourses by lay scholars and then by physicians and scientists. The German and Austrian Jewish scientists and scholars, such as Zollschan, Elkind, Herz, and Sofer (Turda and Weindling 2007) advocated racial eugenics to shape the national image of the Zionist seeking his appropriate space (a land) in the world. In contrast to these scholars, the German-Jewish philosopher and psychologist Moritz Lazarus warned against the implications of claiming to belong to a pure race among other pure races that would shape the Jewish national identity as a separate, mostly inferior, non-European minority (Turda and Weindling 2007).

Arthur Ruppin, head of the Palestine Office of the World Zionist Organization, which purchased lands and established various kinds of settlements, regarded race as a legitimate category and viewed racial differences as significant.[9] Several works have analyzed the racial elements stated in Ruppin's work and the links between the foundation of modern Jewish sociology and the Zionist Movement in Germany. Mitchell Hart's book examines the racial aspects as addressed in Ruppin's work, and devotes an important chapter to "Racial Iconography and Anthropology." Also, Joachim Doron's doctoral dissertation notes Volkisch's influences on Ruppin, and Yaakov Reuveni focuses on the relationship between Ruppin's sociological view and his Zionist outlook.

The concept of race is essential in Ruppin's first book on the Jews, *The Jews of Today*, which appeared in 1904 in German, where he discusses the role of racial characteristics in building the collectiveness of the Jews. Ruppin emphasizes that racial union is the Jews' strongest bond. He claims that the Jews preserved a high level of racial unity throughout the generations. By racial unity he means that race unites the Jews through space and time. Despite geographical distances, the Jews in the West and the East are racially close to each other. His explanation of the continuity of the "Jewish type" for hundreds of years under conditions of geographic dispersion supports his discussion of Jewish nationalism. In 1904 he wrote:

> Many authors deny altogether any unity of race in the Jews. In reality, the Jews, in the course of 3,000 years' history, have assimilated to a small extent certain foreign ethnic elements, though in the mass … the Jews have not only preserved their great natural gifts, but through long process of selection the gifts have become strengthened … For this fact alone the Jews may well claim their right to a separate existence and resist any attempt to absorb them … Mankind of today aims not at uniformity, but at making use of individuality of every nation for the common good.
>
> (Ruppin [1904]1913: 214–218)

This suggested racial model has prompted valuable implications for Ruppin's demographic and social vision. Among these implications, the racial identification offered by Ruppin led to exclusion of some racial groups from the Jewish racial community. Ruppin calls these groups "special types." This category included the Jews of Yemen, Caucasus, and Bukhara, who were racially different from the rest of world Jewry. Another category was termed by Ruppin as "foreign types," which implied an intermixture of Jews with northern Europeans. The racial perspective leads Ruppin to investigate the Jewish people's degree of adaptation to their various environments (Ruppin 1934, vol. I: 7, 14–19). It is implied by Ruppin's discussion that the Jews have the right to exist as a separate *national* unit. Ruppin's nationalist-Zionist view and his view of race are closely connected and were highlighted within the debate over assimilation. The Jews' disappearance as a sociological unit resulted from the Jewish experience of racial assimilation. The aim of Ruppin's warning against racial assimilation was embedded in his Zionist-national vision. Racial uniqueness is a key to national uniqueness and therefore Ruppin believes that there are unavoidable racial or biological differences that do not allow for racial assimilation of one race by the other. Thus, the spatial separation of races coincident with Ruppin's claim for Jewish national segregation was alleged necessary to protect Zionism from racial degeneration.

Taking into account the progeny of interracial marriages, Ruppin supported the idea that such marriages would lead to offspring that are physically and mentally inferior. Here too, Ruppin remained consistent with his deterministic racial theory. He proposed that interracial offspring harm the racial quality of the Jewish race and result in mediocrity.

Israel: eugenics and national redemption

The discretion of the racist form of eugenics and the emergence of more legitimate forms of eugenics such as productivism were also addressed by the Jewish Labor movement. The Social Democratic and Labor movements' goal of preserving their people's-home ethno-national identity became a part of the Labor Party agenda during the 1930s.

The history of the Israeli Labor Party, which has recently been subject to overwhelming revisionist historiography provided by Zeev Sternhell, paved the way to a heated debate by asserting that socialism was at the service of nationalism and the construction of the nation, and that left-wing Zionism made political use of socialism as a rhetorical means of legitimating the national project of creating a Jewish state (Sternhell 1998). Sternhell views the leaders of Mapai, the former Israeli Labor Party, as detached from the idea of the socialist and classless society, universalism, and equality. Yet, he argues, they were able to use the guise of universalism to provide different levels of benefits to different social groups, often European immigrants, at the expense of Arabs and Asian and African immigrants. Similar to Sweden, there was a symbiosis between the Central Union and the Labor Socialist Party. A nationalist-productivist synthesis is the logical ideological conclusion of that blend. This synthesis was complemented at the political level.

The coalition of Mapai, including *Achdut Ha'avodah* (the Unity of Labor Party), the middle-class sector, and the National Religious Movement, addressed the need for creating a melting pot. The Zionist process of nationalizing the state put extreme emphasis on state-building efforts that included the elimination of the foreignness of immigrants (Migdal 2001). Additionally, in order to advance a democratic "republican" system, it was necessary to draw clear prioritization of rights of different ethnic communities. The result was that the western, Ashkenazi Jews were considered the ideal type and Sephardic Jews had to meet the western standards (Barzilai 2003; Kimmerling and Migdal 2002; Matmon 1933; Peled 1992).

Investigation of the nascent eugenics ideas during the 1930s–1940s provides an unusual perspective from which to analyze the project of social engineering that underlay the campaign to settle Jews on the land. During the 1930s the Zionist national productivist eugenics was reflected in negative eugenics demanding the identification and exclusion of "inferior" "degenerate" people who threatened the security of race and state including selective (exclusionary) immigration, while positive eugenic solutions were offered by the prominent Zionists such as Max Nordau, Y. Rubin, I. Rubin, and Fritz Shimon Bodenheimer to increase the Jewish birth rate and Jewish productivism.

The population debate in Israel was shaped by the unique historical circumstances of the Diaspora. The Israeli Zionist movement strove to motivate the Jews of the Diaspora, who lacked any cultural bond with any territory, and to settle them in a populated territory that was already controlled by others. Zionism thus entered into an inevitable conflict with the Palestinian national movement. The Zionists did not hold a hateful racist approach towards Arabs, believing instead that a developed Zionist country would benefit Arabs as well. However, despite their intentions, defining Zionism in ethno-national terms prevented any possibility of communal existence for the mutual benefit of Jews and Palestinians (Peled 1992).

For example, one can view Ruppin's ideas by the late 1920s and 1930s as a eugenic solution towards the Arab-Jewish national question. Ruppin, the Jewish-German founder of academic Jewish demography and one of the founders of "Brit Shalom," fostered Jewish-Arab understanding that aspired to establish the

Jewish national home on the basis of full equality for Jews and Arabs in Palestine, as a binational state. However, due to the Arab riots and uprisings in 1929 and 1936, and thereafter, he came to the conclusion that the time was not yet ripe for fruitful Arab-Jewish negotiations, in which each nation would receive its due. Instead he believed it necessary to strengthen the Jewish position in the country both economically and politically in order to create a situation in which both sides would be willing to cooperate to their mutual benefit. Ruppin came to a eugenics approach to Zionist nationalism by supporting the adoption of a selective policy for immigration to Eretz Israel (Ruppin 1936: 66–70). Ruppin introduced his thesis in explicit national-eugenic terms:

> It would of course be preferable if only strong and healthy persons came to settle in Palestine, so that we would be assured of a strong and healthy succeeding generation. Unfortunately this greatly desired objective cannot be implemented with such generalized simplicity, as the concepts "strong" and "healthy" are not particularly clear. Weak individuals may in due time become strong, for in many professions, the condition of physical capacity and good health is believed to be less significant than mental and spiritual make-up; and, finally, in many cases a person's constitution has no relevance to reproductive capacity to produce healthy children. On these grounds we shall have to limit ourselves to the physical appurtenances of immigrants, and reject those individuals who are dangerous by virtue of some infectious disease (syphilis, advanced tuberculosis, etc.) or who are likely to become public charges (the mentally deranged, epileptics, and those who are prevented by sickness from making a living).
>
> (Ruppin 1936: 78)

Quality was considered just as, if not more, important than quantity. In "The selection of the fittest," Ruppin recommended very specific measures to "raise the level" of the immigrant. The elimination of such persons should be taken care of by the doctors of the Palestine Immigration Offices in the ports of sailing; but a second examination should take place in Palestine in the ports of arrival (Morris-Reich 2006; Ruppin 1936: 67, 78).

During the 1940s and especially with the emergence of the State of Israel, a new mass immigration to Israel, composed of Middle Eastern Jews, became of great concern to the eugenic-based makeup of Israeli citizenry. Following a census of the Jewish mentally ill in 1936, conducted by prolific Zionist neurologist Lipman Halpern, a Berlin doctor who was the founder of the Department of Neurology of the Rothschild-Hadassah University Hospital in Jerusalem, and one of the most prolific writers on the subject during this time, it was estimated that a total of 726 mentally ill Jewish patients were identified in Palestine in 1936, compared to 249 Jewish patients identified in the 1931 census (Eitan 1937; Halpern 1937). This increase was explained, in part, by growth in the Jewish population. Halpern's diagnostic categories on mental illness for Palestine were stratified along the lines of gender, age, and economic impact rather than on racial characteristics. The fact that the census did not consider people over age 60, but only aged 15–30, showed a clear preference for "productivist" measurement of Zionist immigrants and their use in the labor market (Eitan 1937; Halpern 1937). Thus the focus of Halpern's census reveals concerns about the inability of young immigrants with mental illness to work. Since productive work on the land (i.e., farming) was considered the ideal way to devote oneself to the Zionist movement, and people aged 15–30 were considered the most desirable workers, inability to do such labor was considered a serious problem (Eitan 1937). Another important category was the ethnic characteristics of the patients. Eighty-five percent of all mental illness was found among Ashkenazi Jews, while Sephardic and other "oriental" Jewish communities (those of Middle Eastern origin) each exhibited less than 10 percent (Halpern 1937). Halpern believed that different rates of morbidity among Jews could be eugenically explained by the existing connection between psychoses and progressive civilization: "This connection is frequently shown only indirectly through injury to the embryo ... no one becomes insane ... without a predisposition to madness." Increasing civilization, with its elevated number of nervous disorders, indirectly influenced "the deterioration of the hereditary proportion in the next generation" (Mills 1933: 230). Moreover, the Arab/Jewish statistics comparisons show that Jews still exhibited a higher rate of mental illness than other populations in Palestine for similar years (Halpern 1938). These figures emphasized that disability should be considered a national problem and stood against the Zionist ideal of labor as the primary means to achieve a new Zionist homeland in the world of "normal" nations, even as that society was "in the making" (Sufian 2007).

Thus, the debate over the growing numbers of mentally ill within the Jewish people was viewed by Zionist medical professionals and policymakers as a policy problem that should be solved by negative eugenic practices. One of the eugenic solutions was exclusionary immigration (Zalashik and Davidovitch 2009). The Zionist Organization followed the British Mandatory Government (BMG)'s Immigration Ordinance of 1925, which stated that individuals who were identified as mentally ill, physically ill with contagious diseases (e.g., syphilis, leprosy, TB) or epilepsy, lunacy, or idiocy, were excluded from entering Palestine (Baynton 2005; Davidovitch and Shvarts 2005: 411; Shvarts *et al.* 2005). The basis for exclusion was national-productivist, principally to maximize the labor of the fledging proto-state, and through labor, enable spiritual and physical renewal and elimination of those Jews who were unable to contribute to the Zionist national-labor project (Shafir 1989).

The selective immigration process of the Jewish Agency (JA), managed by the Jewish Labor movement, was based on several fixed categories: Category A consisted of persons of independent means (capitalists whose number was unlimited); Category B comprised students and persons of religious occupations whose maintenance was assured; Category C covered people with a definite prospect for employment (laborers whose number depended on the "economic absorptive capacity" of the land and therefore continually changed); and Category D, made up of dependents (women and children) or permanent residents.[10]

Indeed, the Jewish Agency regulated the stream of immigration to encourage the entry of socially productive elements into the Jewish homeland. The first phase of this process included Zionist training camps and medical tests conducted in Europe (Halamish 2006: 60). Every immigrant received a health card that included his medical tests' results performed at the point of departure. Medical inspection was performed once again upon the immigrant's arrival. Immigration certificates were usually given to men between the ages of 18 and 35 who were capable of doing physical labor or had a specific occupation (Sufian 2007). However, despite attempts to control the quality of immigrants coming to Palestine, some people managed to arrive and pass the medical selection process with an illness that manifested itself after immigration (Davidovitch and Shvarts 2005: 419). Expenditure on caring for Jewish mentally ill patients in 1936 totaled 23,000 Palestinian lira (estimated as US$2,584,995 in 2010 terms)[11] (Halpern 1937). The Yishuv expended over 50,000 Palestinian lira in 1941 to maintain its mental health institutions (estimated as US$5,619,554.30 in 2010 terms).[12] Thus, the general shortage of Jewish mental hospitals and health care workers in Palestine made policymakers and health practitioners believe that only exclusionary immigration would reduce the expansion of the country's feeble social margins (Davidovitch and Shvarts 2005: 412).

In order to combat that welfare burden, hundreds of immigrants with disabilities were returned to their country of origin following the guidelines of two of the labor organizations—the Jewish Agency and the Secretariat for Health Matters of the Jewish National Committee in Israel (Davidovitch and Shvarts 2005: 419; Shvarts *et al.* 2005: 10). In 1937, Chaim Yassky (Director of the Hadassah Medical Organization) wrote in the *Bulletin on Social Welfare in Palestine of the General Council of the Jewish Community*, that mentally ill Jews who could not pay their repatriation to Europe were "causing a continual burden on the public and on its social institutions" (Yassky 1937).

Therefore, repatriation policy was funded by another labor institute—the Social Welfare Department of the *Vaad Leumi* (the National Council), which provided 226,210 Palestinian lira. Indeed, the Fund was responsible for sending back tens of chronically ill people, giving 39 people money and contributing to their ship expenses, while it assisted 63 people in reducing their ship expenses only (Yassky 1937).

As in the Swedish case, negative eugenic practices were accompanied by positive ones. Dr. Max Nordau, Theodor Herzl's colleague, had a profound role in this construction. According to Nordau, involving productive Jews in physical labor would help eliminate a root cause of European anti-Semitism, which considered Jews to be parasitic people, and would strengthen efforts to place Jews on Jewish agricultural settlements.[13] Berl Katznelson, one of the intellectual founders of Labor Zionism, regarded Shabbat (the weekly Sabbath or day of rest in Judaism) as a national productivist value rather than a religious one. He claimed that Shabbat was "the solid pillar of Hebrew civilization and the first social achievement of working man in human annals ... I don't need political deals or rabbinical commands to uphold it." Following that, it seems that

Shabbat was a eugenic condition for the survival of the Jews that should be incorporated within the labor movement agenda (Katznelson 1944: 54, 148–149).

In her study, Stoler-Liss (2003) offered evidence of the introduction of "consultation stations," modeled after advice centers for couples who planned to marry and have children, by the Zionist movement back in the 1930s. According to Stoler-Liss, these stations were particularly aimed at family counseling to immigrant populations. These counseling stations were meant to encourage couples to produce offspring to protect a cultural-biological heritage. Dr. Israel Rubin, an educator and literary critic, observed mixed marriages among various Jewish ethnic groups. "Here the 'ingathering of the exiles' in Palestine makes possible 'mixed marriages' not between Jews and non-Jews, but between Jews and other Jews ... Doesn't this in itself contain the hope of eugenic salvation to a great extent?" (Rubin 1933: 89).

These practices focused on stimulating middle-class fertility; however, frequently there were irreconcilable controversies about the aim of the state intervention and social welfare—whether to prefer quality over quantity of Jewish reproduction. Proponents of positive eugenics quickly rose in government service (Stoler-Liss and Shvarts 2004). Dr. Joseph Meir, a prominent eugenicist, who served for 30 years as the head of the Kupat Holim Clalit (General Sick Fund), the largest and first health insurance institution funded by the nascent government, called for the initiation of a single state agency with the goal of eliminating degeneracy and the irresponsible behavior of young couples for bearing unhealthy children (Stoler-Liss 2003). As a eugenic practitioner, Meir harshly criticized the reproduction financial gain of 100 *lirot* that Ben-Gurion promised to every mother who gave birth to 10 children. Meir's critique was against the "universality" of the proposal, which included poor families:

> We have no interest in the 10th child or even in the seventh in poor families from the East. In today's reality we should pray frequently for a second child in a family that is a part of the intelligentsia. The poor classes of the population must not be instructed to have many children, but rather restricted.
>
> (Meir 1952)

The spirit of ethnic republicanism, as defined by Yoav Peled, also manifested itself in laws such as the 1949 Defense Force Service Law, which promoted the discharge of married mothers or married women from military service in order to encourage reproduction (Yuval-Davis 1997). This law also provided loan subsidization and increasing pensions to families with four or more children. The basic idea was *not* to counteract what, in the 1990s, was considered an Arab demographic threat, but instead to promote the reproduction of the fittest—western Jews and Israel-born *sabras*. Although the Zionist solutions luckily did not come close to those of the Swedish sterilization policies, managing the Jewish mentally ill in Palestine broadly fit into the eugenic framework of preventing the "problem of unproductive people" through immigration policy or eliminating it by means of asylum or repatriation. The eugenic practice was conceived as a punitive attitude towards lack of productivity. In Sweden, the "Tartar" and the feebleminded were accused of their "productivity incapacity," while in the Zionist case, Arabs were accused of lacking such capacity.

In both Israel and Sweden eugenics addressed the ideology of productivist welfare planning, but its position within the political system was different. The scientific community readily mobilized itself for the national task, as Porter called "the politics of expertise." In Sweden, as Porter (1996) argues, the state supported and funded institutional academics and influential health and genetics professionals who engaged in eugenics and provided them with increased access to the political process. The process of eugenic institutionalization in the Swedish academy, which took place before World War I, provided a powerful platform for eugenicists to become influential actors in political decision-making and policy formation. These opportunities were strengthened by the social democratic movement's claim to rationalistic modernization of Swedish society. Both medical professionals and anthropologists had an autocratic and elitist perception of their role, usurping the authority to dictate what is best for the society. This was also the case in the pre-Israel state; however, the scientific community did not achieve the same level of state funding and support as that enjoyed in Sweden. In other words, the aspirations of Jewish eugenicists and health professionals were greatly comparable to those of eugenic reformers in Sweden before World War II, but they lacked the power and the opportunities to fulfill them in the same way.

An example of the significant role of the scientific community can be drawn from the way both the Zionist Labor movement and Swedish Social Democrats followed the anthropological crusade against racism, shifting the subject matter of eugenics from the biological to national productivist. This shift is reflected in the social democratic discourse, which exchanges the term "population" for "race." In the Zionist case, the need to legitimize secular construction to Zionist nationalism while avoiding racial stereotyping smoothed the way to a productivist type of eugenics.

Denmark: the social democratic rationale for sterilization laws

As remarked by Nils Roll-Hansen, introduction of sterilization laws in the Nordic countries in the 1930s was carried by the ideology of reform eugenics, at least on the expert side (Roll-Hansen 1996: 262). As in other Nordic countries, Denmark is a national society whose survival depends on its productionist character.

Slightly different from other Nordic countries, Denmark was more continental and relaxed with respect to the Nordic idea. While Finland had a divided population of Finns considered to be Baltic, and Swedes considered Nordic, Denmark is basically Nordic, so there is no debate about "who we are." In technical terms there are very small differences between Denmark and Sweden in the structure of state and society. But nationalism in Denmark is historically and culturally grounded, whereas Swedish nationalism is linked to the dramatic development of the economy after World War II.[14]

Denmark is proud of its socio-economic development, which is an integral part of its national spirit; at the same time it is a country that believes in the necessity of promoting cultural homogeneity.

This does not mean specifically that Denmark is an ethno-national country; however, it expects newcomers to adapt themselves to their new society. This approach does not come out of a sense of superiority, and the proof of it is that Denmark never had the desire to be a great power. It even assumed with pride its defeats and never thought in terms of "terra irredenta." Its fleet and Norway were lost in the Napoleonic wars, while the areas of Holstein and Schleswig were ceded to Prussia after the Second Schleswig War (1864–1866).

Scholars such as Mouritsen stress the concept of *Medborgerskab*, which means membership in society, equal sharing of the same welfare-state goods, and being "one of us" (Mouritsen 2009). Galal adds the concepts of responsibility, maturity, and contribution to social and political life (Galal 2006). The concept differs from *statsborgerskab*, which denotes legal nationality on the one hand, and from *indfødsret*, which literally means the right to be "native born." Today the concept connotes a discretionary exclusive membership. It implies that entrance into the community is possible but difficult. Uncritical or unreconstructed histories of national identity shape citizenship and integration policies, whether by projecting old style historical high culture, or strong and undivided loyalty, or linguistic perfection, as the condition of full national membership. In more senses than one, Denmark could be considered one of these countries in which the conditions of assimilation into the nation is made so difficult as to be considered ethno-nationalist. Interestingly, differently from the south of the continent, it was difficult to find in Denmark a synthesis of cultural pessimism, extreme exaltation of "Nordic" ideals combined with political reaction (Hansen 1996: 12). That means that rather than developing reactionary nationalist ideology, Denmark embarked on developing a national welfare community that does not fall far from that of Sweden.

The advent of the Social Democrats to power in the early 1920s in both countries gave them the opportunity to implement the social policy reforms that they had advocated. After World War I, the Social Democrats gained ascendancy in Denmark, without major conflicts with traditional parties of the center and right. This set the underlying basis allowing for enacting a number of reform laws that gave rise to the Danish welfare state. At the same time, that was a perfect time for enacting eugenic laws, precisely in order to preserve the welfare-state harmony.

Thus, it is not strange that Denmark was the first Nordic country to introduce national legislation on eugenic sterilization, especially if we heed the fact that myths and beliefs on degeneration had been spread throughout Denmark since the end of the nineteenth century.

Agustin Morel, a French psychiatrist, gave scientific legitimacy to the idea of hereditary determinism. The concept implied that social ills such as drinking and sexual excesses were hereditary. As with most Darwinian issues, such a survival of the fittest was hardly used in regard to human beings; issues related to evolution in general or "descent with modification" were popular among progressive liberals.

Of special mention is J.B. Haycraft's *Darwinism and Social Improvement* (1894), the main thesis of which was that the most valuable parts of society reproduced at the lowest level, while the "inferior" part reproduced at the highest level. As a contrast to this negative perspective, it could be expected that eugenicist intellectuals would portray an ideal worldview in which Denmark would fit. However, progressive intellectuals such as Gustav Bang or Wilhelm Johannsen, one of the principal architects of the new Mendelian genetics in Denmark, were against all attempts to favor the propagation of an "ideal" view of the members of humanity. Johansen, for example, suggested that there are a variety of types of humanity; therefore there was not an ideal type. At the same time he claimed that any society should impede the procreation of flawed genotypes.

At the institutional level, early in the twentieth century, those most interested in providing eugenic solutions to social problems were the members of philanthropist and social institutions.

The Danish institutions for the mentally retarded formed a small close-knit community early in the century. These were philanthropic organizations that the state began to support financially (Hansen 1996: 16). However, before state intervention, for a long period of time the mentioned philanthropic institutions were like feudal systems led by families. The Keller family was a case in point. Johan Keller was a philanthropic clergyman who established the Keller Foundations for the Mentally Retarded. Due to the personal connections of this family, his influence on the bureaucracy and politicians was immediate and quite decisive.

However, the interesting development is that these types of philanthropic institutions based on a personal approach became slowly secularized and professionalized. Johan Keller's son, the physician Christian Keller, symbolized the shift from philanthropy to professionalization. Indeed, he was one of the professionals who were appointed in 1924 to sit in the Experts Commission, which was to examine to what extent any interference with the personal liberty of degenerative persons (with or without their consent) would be justified as a preventive measure, with regard to social reasons and racial hygiene.

Keller portrayed the voice of a great majority of physicians in Denmark with regard to the classification and care of the mentally unhealthy. He was in favor of sterilization and even castration of patients, in order to ensure that those with uncontrolled sexual desire will no longer pose a moral or racial threat to the community.[15] However, more striking was the assumption that in order that health institutions would achieve a high degree of efficiency there was a need to reduce the number of the needy. From a utilitarian perspective, the surgical *solution* offered advantages to institutions that were overcrowded.

The questions for experts, for example, were how to treat mentally retarded people; whether patients should give their consent to sterilization and what to do with sexual abusers, etc. Another question to be resolved was whether sterilization measures contribute to the efficiency of health institutions.

Both the negative and positive sides of eugenics were important to progressive technocrats, and, as we shall see, Social Democrats felt most comfortable with the modernist technocratic approach to social health.

Indeed, the "productivist" or "efficiency approach" suited Danish social democrats. K.K. Steincke, for example, was a clear representative of this trend of "new administrators," members of the Social Democratic Party, the party of uncouth trade unionists (Hansen 1996: 27).

In 1920 Steincke produced a reformist manuscript for social health. A great part of the manuscript was dedicated to eugenics. In a non-sophisticated way, he promoted the idea that the fittest, namely the most intelligent part of the population, should produce as many progeny as reasonably possible. Steincke advanced an approach characterized by a seemingly difficult synthesis of health policies and eugenics. The main idea was to protect and feed the feeble as long as they did not reproduce.

However, a shift in this perspective was perceived in 1924, when the Social Democrats took power. Steincke, then secretary of justice, succeeded in putting together a commission that was to consider policies of sterilization of mentally retarded as well as castration of certain groups of sexual offenders (Hansen 1996: 32). The commission was composed of legal experts, physicians, and scientists. In 1926 the commission issued a report, *Social measures toward degeneratively predisposed individuals*. The concept "degenerative" was all-inclusive, and especially stressed the mentally retarded.

However, the most important point advanced by the commission was that it granted an extreme power of decision to technocrats. As in other Nordic countries, the role of experts, physicians, social reformers, legal experts, etc., was of central importance when we come to the question of how eugenic ideas should be applied.

Those technocrats were not racists, and have never promoted any kind of legislation directed to "racial improvement." Since they were sensitive to Nazi eugenic legislation, they stressed their difference with Nazism by enhancing the voluntary character of sterilization in their proposals, as opposed to the enforced ones enacted by the Nazis. However, despite their original effort to distance themselves from Nazi-enforced legislation, in reality the commission recommended, similarly to the Nazis, to sterilize certain groups who were incapable of raising and educating their progeny. The reason was that individuals considered to be mentally ill had no right to make a valid legal decision.

The tension between practical measures and morality was decided in terms of efficiency measures, despite the fact that the law resembled Nazi legislation. To be precise, the German law of sterilization was more specific with regard to the "potential people" to be sterilized, such as hereditary epileptics, hereditary deaf, hereditary blind, and alcoholics. However, although the Danish laws of 1929, 1934, and 1935 did not display such a level of specification, still, the recommendations of the Danish medico-legal council of 1934 referred to a similar list. Indeed, the Danish Mentally Handicapped Act of 1934 also promoted sterilization of psychopaths and habitual criminals, and also of the mentally retarded. In other words, the distinctions were difficult to be drawn at this level.

However, the Danish did not stop at that point. A revision of the marriage law also took place in 1938. Steincke, who was at that time minister of Health and Welfare, proposed a prohibition of marriage to include the hereditary blind and deaf.

In 1939 Myrdal's inspired law of abortion was approved. The law determined that a woman could be sterilized if she was genetically afflicted. It also determined that an abortion could be performed if it was accepted that children might suffer from mental illness. In sum, with the support of progressive elements in Danish society, the number of sterilized people augmented considerably. From 1929 to 1935, 108 people were sterilized, but in the next five years the number increased to 1380. From 1940 to 1945, another 2120 were sterilized, and until 1950 another 2000 people suffered the same process—most of them women and the mentally retarded.

National consensus on a productivist eugenic welfare identity

At this point we should remark that if Danish Social Democrats were the political forerunners of the eugenic solution to social problems, it is impossible to deny that there was a Danish national consensus regarding the conviction that eugenic measures were an integral part of programs of social and health reform. That consensus encompassed progressive politicians and scholars as well as members of the Lutheran Church.

It must be remarked that the first sterilization law was passed in 1929, when the reactionary Agrarian Party replaced the Social Democrats in power. However, Social Democrats, especially K.K. Steincke and Vilhelm Rasmusen, were the most extreme supporters of the law. They supported the law because of its eugenic progressive character, such as the potential benefit of the law for the whole of the society. However, support for the law did not only come from the political party. In civil society we could find arguments in favor of eugenics stemming from people associated with women's liberation. Indeed, progressive women such as Thit Jensen, a famous Danish feminist, endorsed eugenic arguments, for example, because she assumed it would help in birth control. Indeed, this approach represented a sign of women's liberation and emancipation from men's control.

Jonathan Leunbach, a physician and cofounder of the *Magnus Hirschfeld League for Sexual Reform* and a founder of the *Sexual Reform Party* associated with the Communist Party, also supported eugenics because he saw it connected to birth control and sexual emancipation. However, this support coming from socialist and feminist intellectuals suited perfectly well a "whole Danish state of mind," or what could be defined as the Danish collective identity composed of free rational individuals. Indeed, to most Danish people, rational human beings should control both reproduction and the quality of the newborn.[16]

In the book *Arv og Race* (Heredity and Race) published in 1934, eminent authorities (Ojwind Winge, an expert on genetics; Oluf Thomsen, an expert on human genetics; August Goll, a legal expert; and also the noted K.K. Steincke) presented articles on genetics, eugenics, and social policy, and on race and racial biology, which constituted a celebration of the mentioned Danish consensus (Hansen 1996: 44).

The consensus, as noted, relied on the conviction that eugenics provides solutions to social problems. However, in addition to that, and in response to the growing suspicion that Danish eugenics resembled that of the Nazis, the Danish remarked that racist and anti-Semitic ideas are wrong. However, when it comes to eugenics and especially eugenic German laws, the tone of Danish eugenic intellectuals changed. Authors such as August Goll, for example, maintained that the German law on eugenics should be considered positively. In other words, Goll, who was anti-racist and at the same time advocated sterilization laws, strove to disentangle eugenics from racism. If eugenics is disentangled from racism, then we can assume that although the Danish law of sterilization did not differ radically from the German sterilization law of 1933 (Law Concerning the Prevention of Hereditary Afflicted Progeny), still and despite that, Denmark could not be compared to Nazi Germany (Kemp 1933). Denmark, at least theoretically, thought in productivist terms, while for Nazi Germany race trumped productivity.

As we can grasp from the previous section, eugenics was portrayed as a neutral technical field of social policy, which could be used by reactionary racists as well as by progressive people. That is the reason why Danish supporters of eugenics could have condemned Nazi racism while not condemning Nazi eugenics.

This comes as no surprise, although at the same time it raises questions on the exclusionist character of Danish collective identity.

What is clear, though, is that eugenics had a deep history in Denmark and was especially related to the establishment of a welfare society. With the passing of years that part of the history was ignored. There was no pride in that epoch; however, as we have noted, there was no sense of guilt. No public debate when the sterilization law was revised in 1967, and there was no debate when the amniocentesis test was introduced in 1970.

In her study of the Danish sterilization practice over the period 1920–1967, Lene Koch suggested that applications for sterilization in Denmark were based on purely eugenic indications, or purely social, or they were both eugenic *and* social indications at once. Any reference to inheritable conditions for sterilization was assumed to be a eugenic-based application. These latter were not numerous. Applications with purely social indications approached 50 percent of the total. The question is why there were so few applications for sterilizations for eugenic reasons. According to Koch, the role of eugenics was so pervasive and appeared to be so evident to those involved, that mentioning it was not always required (Koch 2000: 334). This is the reason why even in the parliamentary debates that led to a ban of forced sterilization, the parliament members did not connect forced sterilizations with eugenics. While forced sterilization was condemned, the eugenic indicator was maintained for mental normality and eugenic considerations were taken into account. Voluntary sterilization and abortion were legal if compatible with personal freedom.

It comes as no surprise that this benign setting for eugenics constituted a fertile ground for foreign investment in human genetic research. Indeed, besides the Danish "mentality" with regard to eugenics, Denmark enjoyed a high level of social organizations, civic records, and high-level professionals, which made it a country with huge opportunities for investment in human genetics research.

This convinced the Rockefeller Foundation to found in 1938 an Institute of Human Genetics, which was immediately handed to Copenhagen University. Tage Kemp, a scholar who from 1933 had been writing regularly on the subject of human heredity and eugenics, became its first director. Genetic registration was the central activity of the new institution, together with family research and also experimental research into pathological genetics.

The Institute not only contributed in research but in providing advice on marriages, on sterilization, and other things related to human genetics.

Tage Kemp continued his activities as a popular lecturer and ideologue of soft negative eugenics, whose basic principles according to Kemp appeared in an article in the *Eugenic Review* in 1957.

Kemp claimed that "To judge from theoretical calculations, the negative selection, which can be effected by moderate genetic-hygienic measures may probably cause a considerable fall in the incidence of the hereditary diseases in the population" (Kemp 1957: 18). Kemp praises the development in Denmark as a precursor of genetic hygiene, and continues by saying that "Genetic hygiene is here based on the principle of voluntariness, and the population has learned to understand its purpose. Democratic social conditions prevent misuse and secure that due consideration is paid to personal liberty" (1957: 18).

This approach fit quite well with the Danish working class cultural policies generated by Social Democrats, which gave expression to the goal of crafting a healthy national population regardless of ethnic or religious origins.

Differently from Sweden where, until after World War II, there was no " 'cultural policy' as a separate administrative area," (Larsson 2003: 182) in Denmark there was a debate about the creation of an international workers' culture. This initiative ranked high on the cultural and political agenda of the Social Democrats, right after acceding to power in 1924. In 1932, for example, Julius Bomholt, who authored a book titled *Arbejderkultur* (Workers' Culture), was commissioned to draft a cultural policy by the new social democratic government (Tawat 2011: 121).

This worker's culture entailing "mutual solidarity, pride in the job, a special jargon, worker's song and the trade unions" was pitted against the bourgeois culture that relies on opera, ballet, etc., considered to be damaging activities (Duelund 2003: 35–36).

However, the interesting thing is that this international culture fit with what could be defined as a national culture based on one people, folk, and language.[17] Social Democrats sided with Thorvald Stauning's *Danmark for Folket* (Denmark for the People) program. This latter program represented a "middle of the road" strategy conceived to mobilize all social classes and parties around a social welfare project. In 1953, the new cultural policy orientation materialized with the publication of the report *Mennesket i Centrum. Bidrag til en aktiv kulturpolitik* (Focus on the Individual Citizen. Contribution to an Active Cultural Policy).

Indeed, the design of a worker's national culture is important, because it signals and stresses the character of Danish collective identity as a *productivist* identity. The state had to lend a helping hand if the cultural objective of free and equal opportunity was to be realized (Duelund 2003: 39). The policy outcome, as Duelund puts it, was that "culture was now perceived and defined, first and foremost, as a national phenomenon" (Duelund 2003: 41).

In more senses than one, that national culture also applied to foreigners who could be included within the 1933 Danish Law of public welfare. As part of Thorvald Stauning's welfare-state project and the reform implemented by his Social Affairs Minister Karl Kristian Steincke, the social welfare regime was reformed in 1933.

Four main laws were introduced: The first law, *Lov om offentlig Forsorg* (public welfare), replaced the previous Poor Law by new public welfare legislation. The second law, *Lov om Folkeforsikring* (public insurance), dealt with medical insurance, sickness funds, disability insurance, childcare, and retirement pensions. The third law, *Lov om Arbejdsulykkesforsikring* (work-related injuries), regulated work conditions. The fourth law, *Lov om Arbejdsanvisning og Arbejdsløshedsforsikring*, created an unemployment allowance (Tawat 2011: 105).

Basically, foreigners could no longer be stripped of citizens' rights when they became recipients of social aid. At the same time, foreigners had to naturalize and the numbers of naturalized people had not been high since 1935. Applicants for citizenship should learn to speak Danish and share local beliefs and practices; in a sense, be culturally assimilated (Holm 2006: 74).

The interesting thing is that criteria did not change after the German occupation. Before occupation, Denmark had accepted the J pass, which allowed stopping Jews fleeing from Germany. However, at the same time, ethnic Danes, foreigners, and Jews, whether Danish or not, continued their normal life to the extent that foreigners were still eligible for social assistance and had access to jobs. This cannot obscure the fact that in August 1943, Jews were arrested by the Nazis and sent to the concentration camp at Theresienstadt. After that act, however, the Danish people engaged with their Jews and collaborated in evacuating 7000 of them to Sweden. The support for Jews was so strong that, according to a widespread legend, King Christian X himself wore a yellow star as a sign of solidarity.

While occupation was an extraordinary situation, difficult to evaluate, it gives a glance into what could be defined as a strong civic identity that included all those who wanted to integrate.

This cannot be defined as a liberal open identity, since its exclusionist character was determined by a productivist eugenic ideology and practice. However, for all practical purposes, the Danish criteria of nationhood and eugenics were portrayed by Danish technocrats as productivist rather than racist.

That conviction set the underlying basis for a difficult conceptualization and even confusing political situations. Denmark never condemned Nazi eugenic policies. Even when Denmark regretted the enacting of the Mentally Handicapped Act of 1934, that was short of condemning the Nazi law of sterilization.

Even when in the early 1930s the focus on eugenics was replaced by new attention in human genetics, no criticism of eugenic policies could be heard.

The final question, however, is what of this eugenic ideological tradition that has become part of a collective identity remains in present times, when deep liberal and human rights awareness dominate the political and social spheres?

Differently from Sweden, where its eugenic past has triggered a public debate and where since the 1960s an awareness towards diversity was promoted by Social Democrats, Denmark remains an assimilationist country (Hedetoft 2009: 122). Denmark still synthesizes the idea of a welfare-productivist society entangled in a type of assimilationist republicanism with regard to newcomers.[18] There is a logical continuity of those central values, which were first and foremost advanced by Social Democrats.

According to Alf Hedetoft, Denmark articulates ideas that a multiethnic society should not be necessarily multicultural (Hedetoft 2009: 111). Mahama Tawat added that ideas stemming from a national culture, even today, prevented multiculturalism from taking root (Tawat 2011).

As noted, the emerging national culture still has a direct link with eugenics, especially since there is an increasing conviction that demography is more compelling than ever. The question is whether we accept Ulf Hedetoft's argument that Denmark is becoming more open to diversity precisely because of the need for more population, or we assume that precisely the need for more population makes necessary a reconstruction of the eugenic discourse and of national integration policies. One trend seems to be complementary to the other.

Observers might wonder what this owes to the past, and how it is related to Danish understanding of eugenics. In more senses than one, the eugenic project suited a "productivist" concept of exclusion mentioned before. In other words, e.g., nobody should be excluded because of their black skin. However, non-productive, feeble individuals should not be aided to reproduce. Nowadays, because of growing suspicion, rightly or wrongly, that civic integration is failing, the alternative of recreating a productivist society through eugenics is more enticing than ever. As noted, there was no special debate on eugenics in Denmark because the issue was embedded into a national welfare-productivist identity. This assumption is the key to understanding how the new ART programs are also embedded in Denmark's identity.

Finland: eugenics and nation building

The Finnish case can be included within those countries in which eugenics served the purpose of nation building. As we shall see, what started as a particular initiative of the Swedish-speaking minority to "protect itself," became an integral part of the Finnish welfare state, and a tool of national modernization.

Moreover, differently than other Nordic countries, the leading agents of the eugenic movement were not Social Democrats, who were defeated in the Civil War of 1918, but the right-wing White victors of the War. Nonetheless, the striking thing is that although Social Democracy's role in Finnish politics was limited, still Finland was portrayed predominantly as a Lutheran social democracy (Stenius 1997) and the contribution of social democracy in promoting the sterilization laws of 1935 was prominent.

As a mode of introduction, let us stress that the Finns are one of those peoples that are difficult to define. They were ruled by Sweden until 1809 and then by the Russians until their independence in 1917. Their language is related to Hungarian, and to a wide variety of languages of Northern Russian origin and even Mongolian. They do not belong to the East, and according to some scholars such as W.R. Mead they hardly belong to the West, to the extent that some scholars are ready to assume that they are unique, a mystery that is beyond comprehension (Lewis 2005; Mead 1993).

In contrast to similar identity projects of other nations in Europe, the Finnish Identity Project did not revolve around the question of "who are we?" but rather around "who we are not" (Adolf Ivar Arwidsson, a Finnish political journalist, in the first half of the nineteenth century coined the Fennoman credo: *Svenskar äro vi icke längre, ryssar vilja vi icke bli, låt oss alltså bli finnar*—Swedes we are no longer, Russians we do not want to become, let us therefore become Finns). However, to become truly Finns, the Finnish people had to overcome the tension between the two most important language communities of the country: the Finnish majority and the Swedish-speaking minority. According to scholars such as Jouni Häkli, the network of relationships between the Finnish majority and the Swedish-speaking minority has been, for the most part, a "peaceful coexistence" (Häkli 1999: 123). At the same time, it is hard to dismiss the social conflict between a Swedish-speaking minority elite and the Finnish speakers' majority, which has been gaining ascendancy since the early twentieth century. The Swedish-speaking minority occupied the most important administration posts, and was prominent in industry and commerce during the nineteenth century, when Finland was an autonomous Grand Duchy of the Russian Empire. Furthermore, at that time Finland retained the social and legal traditions of Sweden, specifically with regard to family law (Bradley 1999: 175). Hence, social welfare policies have emerged from these cultural traditions that, in turn, have defined practice.

With independence, there were various attempts to craft a Finnish identity, and to this end there were different attempts to establish a national language. It was argued that only a linguistically homogenous people, one speaking the same language, could craft a strong sense of national purpose. The emergence of this idea put the Swedish-speaking population on the defensive (Hietala 1996: 196).

Indeed, if we look for the sociopolitical background behind the appearance of a eugenic movement, we should trace it to the standings of the two languages and collectivities that composed the Finnish nation. If the Fabians in Britain projected their eugenic trend on a social class basis tending to strengthen the middle classes, in Finland the very idea of racial hygiene was first embraced by the Swedish minority.

As scholars such as Marjatta Hietala claim (1985: 517) the "Mongol Finn" idea held by a majority led to eugenics and financial incentives for the minority of Finnish-Swedes to have more children. The first apparent goal was to protect a minority. The second ensuing goal, however, seems to be far reaching. The very idea of a Finnish national identity connected to Europe, rather than to the East, depended first and foremost on the hegemonic role of Swedish-speaking Finnish. While Finnish-speaking nationalists were interested in other Finno-Ugrian peoples, such as Estonians and Hungarians, the Swedish-speaking minority tended to search for links with the Scandinavian countries.

The Swedish-speaking minority thus consider that they should rule over a supposedly inferior race representing a primitive nationalism.[19]

Regardless of the Swedish-speaking minority's belief that they were the Europeans among the Finnish, the question of how Finns should be understood in racial or, more recently, genetic terms has been the source of intense debate for two centuries. Some authors asked themselves what was behind the myth of Finnish "Europeanness." Until the beginning of the twentieth century, anthropologists and geneticists agreed that Finnish language was related to Sámi—and that Finns were Mongoloid. This view remained relatively unchallenged until the beginning of the twentieth century. Since that point and particularly since the 1990s, it has been increasingly argued that Finns should be seen as European (Dutton 2008: 171).

Scholars such as Aira Kemiläinen have argued that Finns have no Eastern dimension at all and are completely European (1998: 228).[20] The question, however, is why this should be important. According to Dutton (2008: 171) this "struggle to be European" has all the makings of a scintillating—and yet dangerous—anthropological myth.

The Europeanization trend was particularly evident after the end of the Civil War. The victory of the Whites during the Civil War was interpreted in Swedish-speaking circles as a victory for Western culture. Becoming European thus implied being ruled by the Swedish-speaking minority or at least adapting to its ways (Dutton 2008).[21]

However, the interesting thing is that the victory of the Whites did not imply revenge against their enemies. The very idea of social integration of the defeated was in the plan of the victors. As can be expected, though, the very process of civic integration was to be advanced according to the paradigm of the nationalist victors. In the young and divided nation, the victors' interests featured the ideal of national unity of a population physically strong, loyal, and patriotic (Satka 1995: 59). The victory of the Whites over the Reds in the Civil War left some 30,000 causalities, mostly Reds. Such being the case, the urgent thing to do was to take care of the upbringing of the children of the Reds. In other words, the winners were especially concerned to integrate the defeated Reds, while the integration of the Reds' children through welfare programs was to create a united nation.

The strategy to perform that task was to strengthen the state by means of establishing a modern public administration that will advance the worldview of the Whites.[22]

Public servants were in a situation that necessitated the production of categories and concepts that extended the institutional scheme of relief of the poor and its interpretative practices as required by the worldview of the winners. Eugenics played an integral role in that worldview, which advanced a conceptual synthesis between Europeanization and social modernity.

The biological sciences seemed to offer solutions to social problems, and in that sense the medical experts contributed to what was supposed to be the modernization of the nation as a whole and not just limited to the Swedish-speaking population.

Since the late 1920s, the scientific discourse of eugenics has been adopted by the large Finnish population. As we shall see, what began as an issue of special concern for the Swedish-speaking physicians, social workers, and population in general, was also adopted by the non-Swedish-speaking Finnish majority. This leads us to the question of what was shared in common by the two populations.

Both of these populations worried about the decreasing population and its qualitative deterioration. It is true that the Swedish-Finns were more focused on maintaining their presence and supremacy in Finland. However, the rest of the Finnish population after independence was especially determined in fighting for world recognition and for the development of a modern nation, a process that translated into the drive to become more European. The operative features of becoming European entailed the rush to be receptive to technological innovation coming from the "developed" modern world and the development of an efficient state bureaucracy. As we shall see, eugenic solutions fit perfectly well in the plans of national modernization crafted by the new bureaucratic elites rising up after the Civil War.

Finnish exceptionalism or Swedish-speaking minority racism?

The *Europeanization* of Finland after Independence was not a difficult task if we remember that Finland was not isolated from European social and intellectual development. Finnish doctors educated in German universities and Helsinski University had strong ties with the German-speaking world. Swedish-speaking Finnish intellectuals were especially connected with Swedish intellectuals and doctors, similarly to doctors of other Nordic countries involved in the international eugenic movement. As can be expected, those most interested in eugenic measures were the Swedish-speaking Finns. In an article entitled "Eugenic work in Swedish Finland," Prof. Ossian Schauman, the first chairman of the Association of Public Health in Swedish-speaking Finland, made clear his plea that Swedish culture should prevail over that of the Finnish.

The first signs of racial eugenics can be traced to 1912 after publication of the reports of the Family Research in Giessen and the Eugenic Conference that took place in London. Expressions of racial differences from the medical perspective were stressed. In the same year, Georg von Wendt in his book *Our Duty to Future Generations: Some Principles of General Eugenics*, made racist claims in the name of eugenics. Prof. Albert Palmberg, in a lecture for the Ekonomiska Samfundet, remarked the social and national consequences of the current birth rate and rates of infant mortality. As others in Nordic countries, Palmberg defined the mentally ill, alcoholics, etc., as degenerate.

The combination of the terms welfare and sterilization were discussed first in the 6th Nordic Conference on the Welfare of the Handicapped in 1912. One of the outstanding lectures was delivered by Dr. Albert Björkman, then the medical superintendent of the Pitkäniemi Central Institution, who suggested that sterilization was necessary in order to prevent increased incidence of mental illnesses, feeblemindedness, and mental degeneracy. Those measures should be applied to normal criminals, sexual criminals, idiots, genuine epileptics, and finally to alcoholics (Hietala 1996: 204). Indeed the vast majority of eugenic supporters were Mendelians who were invested in discovering the roots of antisocial behavior (Portin and Saura 1985: 41–45).

All these elements were taken into consideration with the founding of the first institution of eugenics represented by the Association of Public Health in Swedish-Speaking Finland, founded in 1921. It was described as a Eugenic Society and its mission was to promote public health among the Swedish-speaking minority. It must be stressed, however, that the noted negative aspects of eugenics expressed in sterilization proposals were accompanied by positive proposals.

The Society encouraged mothers to give birth to children of the "Swedish tribe." In order to qualify for an award, mothers were to have given birth to at least four children who were then between 4 and 17 years old, who were physically and mental healthy. Between 1920 and 1928, a total of 601 mothers were awarded special diplomas. Henry Federley, a Finnish professor of eugenics, considered to be a pioneer in the Swedish-minority eugenic movement, remarked that the awards for mothers had a eugenic purpose. It was made clear that the Association of Public Health in Swedish-Speaking Finland would encourage only the best of the population in order to progress. In other words, the program was not conceived for helping poor or large families, but healthy Swedish-speaking ones.

Until the 1920s, the Finnish majority was not involved in the eugenic debate; however, already in the mid-1920s we can perceive that the term eugenics began to be uttered by non-Swedish-speaking officers and the population in general. As we have noted, the idea of crafting a whole inclusive national identity accompanied the idea of Europeanization. Therefore, questions related to the right of the state to manage the number and the quality of the population did not remain limited to the Swedish minority. That issue became an integral principle of nation state building.

As we have noted, the most relevant part of the nation-building strategy was the enhancement of the role of the public service in establishing the version of welfare society emerging with the Whites' victory in the Civil War.

The National Board of Social Affairs, which was part of the government and therefore supported the Whites' worldview, was in charge of smoothing over the gap that had emerged between Finnish people (Satka 1995: 62). The early stages of its work was to single out the objects of poor relief institutions. They were the children and mothers of the defeated Reds. This distinction and classification, which at first glance was to serve the process of integration, was soon universalized. No more Reds and Whites, but the new forms of classification on moral grounds, as either good or bad, loyal or deceitful, became a central organizing principle of the ideology and an institutional form of poor relief. As such, an investigation of people on moral grounds was one of its main functions (Satka 1995: 80). If a person violated the law that the winners considered as sacred (or in the case of illegitimate children, the self-evident moral order of marital sexuality), it followed that the person was investigated as one who belonged to the category of "flawed elements." During the 1930s, the concern for governing led to the desire of "regulating" the flawed elements of society. They had to be subjected to strict discipline instead of caring for them and assisting them. In times of economic recession, the public became anxious about the number of unproductive people.

The increase in criminality as a result of the economically hard times and of the Prohibition Act (which, similar to the United States, prohibited the marketing of alcoholic drinks) led to numerous prosecutions of illegal marketing of spirits and of illegal drunkenness in the public sphere.

In September 1933, some influential groups gathered at the University of Helsinki for a one-week seminar of strategies to prevent criminality. The pioneers included experts in vagrancy, mental illness, and poor relief. Experts in these areas dominated the seminar, in which two leading foreign scholars were invited—August Ley, a professor of criminology from Belgium, and Olof Kinberg, a professor of criminology from Sweden. The Finnish expert in eugenics Herny Federley opened the sessions. In his remarks, he stressed the topic of public order and national governing. The main claim was that until the turn of the century, the Finnish people were considered loyal citizens, but now they had become increasingly criminal. It was necessary to save the Finnish nation.

According to the speakers at the conference, the question of criminality should be also explained as a problem produced by the number of unmarried citizens, and the number of able bodies receiving poor relief treatment.

Brynolf Honkasalo, a professor of law, stressed that the deterioration of the Finnish race was a social fact (1934). The reasons were that the number of genetically defective elements was beginning to outnumber the genetically valuable elements whose rate of birth was rather slow due to contraception. The issue of the negative birth rate, falling below the mortality rate, was portrayed as a whole European issue and if that trend coincided with the proliferation of degeneration, then the prospects for the future were considered awful for European societies.

There was conviction of the need for more political control through more effective scientific conceptual tools to prevent society from moral decadence. The prevention policy, however, needed a reorganization of public service. It needed to become more centralized and effective. The Ministry of Social Affairs thus became more centralized and more dedicated to the regulation of flawed elements.

We must stress that until the 1930s the focus of the debate was negative eugenics with a clear productionist tone. The debate focused on preventing people with unwanted hereditary characteristics from having children and on whether the degenerative type would outnumber the healthy. However, another point became important as well. There was a general anxiety among Finns about the decrease in the population of Finland. That topic was becoming a central issue of debate. The writings of Alva and Gunnar Myrdal became popular among Finnish scientists. Armas Nieminen, a professor in social policy, argued that the stagnation in the increase of population was a symptom of the weakening of the political importance of the nation. Decrease in population in general implies degeneration (Hietala 1996: 238).

To the question of numbers there was an interesting addition, which was related to the types of people that should be encouraged to reproduce for productive reasons. The productionist tone was especially perceived by Vilho Annala, a professor of economics, who explained the consequences of salary reduction in public service. Considering public servants as the elite of society, the author wondered whether civic servants will be induced to keep small families and not procreate because of salary reductions. The author hinted that the human conditions to serve in public office are hereditary, and suggested that the elite of public servants would be followed by their children. This productionist assessment would lead to the idea that public servants should be induced to create larger families. If that represented the positive productivist tone, the negative complementary point was accentuated by Dr. Albert Bjorkman in his address to the Medical Association of Turku in 1925. He noted that the existence of a great number of mentally ill, retarded, and psychological degenerates constitutes the reason for financial losses (Hietala 1996: 218).

Both productionist argumentations were akin to the idea that a strong nation can be shaped by the two communities, Swedish-speaking and Finns, if they rely on the voice of experts. The target of the state was to increase and improve the quality of the national population in general.

Productionist-eugenics sterilization vs. liberal contraception

During the 1930s and 1940s, Finnish physicians actively advocated a law allowing sterilization for eugenic reasons, namely heredity and race hygiene. As noted, one of the reasons for this attention on eugenics was a latent desire to be part of the modern world. Finland adopted eugenics in the same way that it endorsed scientific innovations during the post-independence era. The very idea of becoming a progressive nationalist society lay at the basis of the drive for adopting scientific trends from the modern world.

However, until the 1930s, sterilizations were forbidden by Finnish criminal law, allowing them only if they were used to care for women's health. However, pro-eugenic groups believed that a general law of sterilization, not limited to women's health, offered a more humanitarian way to deal with social issues (Hemminki *et al.* 1997: 1876).

Prof. Allan Serlachius, in a meeting of the Finnish Association of Lawyers and the Medical Association, advocated the idea of preventing retarded people from reproducing. However, he remarked that the measure should not be applied on racist background (Hietala 1996: 217).

The Finnish government appointed a committee to prepare a report analyzing the pros and cons of a law of sterilization for the mentally retarded. As could be expected, the first recommendations of the committee featured profound productionist content. Indeed, the most striking point in the report was the association of the mentally retarded with lower occupations in the job market. The samples the committee picked as examples of candidates for sterilization were unskilled people, elderly women, a man born out of wedlock, and housemaids. The message was clear: there were no candidates for sterilization who were representatives of higher occupations. Theoretically, the flawed conclusion would be that housemaids could not give birth to successful children. This gives a glimpse into how the Finnish society valued public service.

There was public debate on the recommendations by the committee completed in 1929, whose main recommendation was the enactment of a law that demanded the acceptance of the patient by a guardian.

Generally, the Finnish population supported the recommendations of the Committee. The increase of welfare costs, of "unfit" people, and the increase in crime made the case for popular support. An especial support for sterilizations came from the National Board of Health, a group that was already actively involved in the issue of sterilization.

The debate in the parliament finally materialized in the Government Act of 1935. The Act allowed the voluntary sterilization of legally competent persons if there was justified fear that they would produce inferior children or if because of their sexual behavior they would be prone to abuse others.

However, the important point was to come next. Castrations without voluntary acceptance were also allowed, if the National Board of Health decided on it. Candidates included mentally handicapped, mentally ill, and epileptic people. Moreover, epileptics and deaf-mute people were not allowed to marry without being sterilized.

Similarly to Denmark, the Finns considered "idiots," "imbeciles," and the mentally insane, in general, not qualified to make individual decisions, and therefore there was a reason to decide for them. Finally, those guilty of a crime or attempted crime because of sexual drive qualified for compulsory sterilization.

Support for the law came from different social sectors. As in Denmark, special support came from women. However, differently from Denmark, the women supporting sterilization were not only progressive but also right-wing conservative women who were convinced that sexual abusers should be sterilized.

Social Democratic women, through their organ *Toveritar*, encouraged by the experiences and good progress in the United States, Germany, and Sweden, also expressed support for the law of sterilization.

Between 1939 and 1950 only eugenic productivist types of sterilization were registered. The dominant ideological criteria were sexual morality, creation of a unified nation state after a civil war, and the ensuing political conclusion that would be pronatalist policies.

A good citizen was defined as a sober, honest-speaking, and hard-working member of the state. Motherhood and motherliness were self-evident characteristics of an ideal female. Upper classes, the holders of reform discourse, defined themselves as shepherds and guides of the common people. Their duty was to deliver the enlightenment (Satka 2009: 17).

Different institutions were in charge of promoting the national health gospel. *Väestöliitto*, the Population and Family Welfare Federation of Finland (founded in 1941, composed of 31 member organizations), which still in present times works in the fields of social affairs and health, set out the right combination between welfare and nationalism or between a healthy population and a strong nation.

The federation declared that the country needed a population program and large-scale population propaganda, following the example of the awards to mothers who gave birth to four or more children promoted by the Association of Public Health in Swedish-speaking Finland in the 1920s. At the end of the 1930s, a debate on the ideal number of members of a Finnish family took place.

The central part of the state propaganda was directed to mothers. Motherhood was especially important for the nation's continuity (Satka 2009: 238). There is no question that the experience of the Winter War of 1939–1940 raised concerns over the number of Finns lost because of the war. However, the concerns over the quantity and quality of the population surpassed the war experience. It was considered to be a central issue for the survival of a welfare society. The state ideology was clearly in favor of big families and a certain cult of the nineteenth century type of home became a central characteristic of the state. The state encouraged women to have six children, four children at a minimum.

Regarding the values that should be encouraged in the population, those should be related to the countryside (as against the values of urbanization). Special attention was paid to regional settlement in the most remote parts of the country such as Lapland.

Since the 1940s the thinking in Finland headed towards a pronatalistic ideology when the state overtook a reformist welfarist stance (Hemminki *et al.* 1997: 1876). The state offered constant aid to its citizens, promoting the training of vocational and professional experts for health care and social welfare. Teachers and administrators of social work in new mother-child clinics and municipal welfare offices became key actors in the state program of construction of the welfare state (Satka 2009: 239). As we can see, the negative and positivist aspects of eugenics met. The productionist side of negative eugenics (limiting the cost of maintaining the growth of the social margins of society) met the nationalist positivistic side of race hygiene calling for settlement of the country and the growth of productive healthy families.

Productionism and national integration

The support for negative eugenics did not fade away, and another Law of Sterilization was enacted in 1950. The question, however, is: what was the difference between the 1935 Act on sterilization and the new Sterilization and Castration Law of 1950? According to Hemminki *et al.*, the spirit of the new 1950 Sterilization Act was the same as that of the 1935 Act. The background was a perception of a rising crime rate in Finland. Following the Law Relating to Castration (S 1. S 6),

> Any person who, by virtue of a judgment to which legal effect has been given has been found guilty of an offence showing that he is a danger to others by reasons of the violence of his sexual impulses, may be castrated without his consent.
>
> (Törnudd 1996)

However, at other levels, the rationale for castration faded away with the modernization of social life.

Already in the 1950s the abundance of abortion among the healthy population increased pressure for contraceptive sterilization. From the 1950s only medical grounds were the basis of sterilization, and from the 1970s there was a shift from eugenics to contraception as the basis for sterilization. Contraception (the possibility of the woman to prevent births by other methods) and social reasons (women's life circumstances) were introduced as new reasons for sterilization (Hemminki *et al.* 1997: 1877).

A new abortion law was introduced and it stated that if the mother or father of the fetus has inheritable characteristics causing the child to be handicapped, the woman's sterilization was required. Combining abortion and sterilization became very common.

By the 1960s, eugenic-forced sterilizations were not considered appropriate and eugenic thinking lost its appeal. The drop in sterilization rates in the 1970s may be explained by the wider opportunities for abortion.

In short, eugenics and incapable caretaker were legal reasons for sterilization during the whole period, but these rationales were rare after 1970. In other words, the rationale of hereditary, race hygiene, and even national productionist reasons were no longer compelling.

According to Hemminki *et al.*, difficulties in contraception were introduced as a reason for sterilization in 1970, and it was increasingly used until the introduction of the age and parity criteria. Since the late 1980s, practically only the age and parity criteria have been used to justify sterilization (Hemminki *et al.* 1997: 1878). As remarked by Bodil Ekman and Jerker Liljestrand, we can see in the Scandinavian countries in general that from the 1970s the case for sterilization of disadvantaged people was made more difficult, while voluntary sterilization was made generally available.[23]

This approach also had an impact for positive eugenics. There was not always a coincidence between the pronatalistic policies of the government and Finnish women and men who, as a result of the modernization of social life, wanted to regulate their fertility. The government's desire to promote fertility of normal productive people was the reason why physicians did not facilitate means of contraception. Only sterilization was available.

This is an important point for current times, because the government is again promoting pronatalistic policies. This raises the question whether the eugenic thinking of the Finnish nationalist period is being reincorporated nowadays under another label, and what this implies.

We can certainly point out that eugenic thinking was prominent in the nationalist era. The question is whether it is possible to detect a new wave of productionist nationalism in present times.

Indeed, the two waves of nationalism in Finnish history were the national awakening during the nineteenth century, when Finland gained autonomy from Sweden, and later independence; and the second wave, which followed World War II, especially after the achievements during the Winter War and the Continuation War against the Soviet Union.

In recent years, a third wave seems to have appeared: immigration has become one of the most burning questions in Finnish politics. In fact, the attitude toward immigration has undergone a significant change, as it is no longer seen as a mere aid to refugees or people in need, but as a political issue, which is now heavily regulated by laws and criteria, all of which are aimed at helping the host country, rather than the immigrants themselves (Peltoniemi 2011: 1–15). This question fits well into the debate initiated in the 1990s around the Europeanization of Finnish identity. Following this line of argument, the very idea of promoting an increase of the birth rate among the national "European" middle and upper classes becomes a national necessity. That would be a national strategy contrasting with the idea that policies of immigration and integration are necessary for economic growth.

Summing up

This chapter locates a shift from racial eugenics to national-welfare eugenics during the 1930s, when eugenic policies could no longer be understood independent of socio-economic factors. Eugenic strategies were favored by social democratic and labor movements to preserve their people's-home ethno-national identity. This may sound like a paradox since pursuing a purely productionist goal implies bringing welfare for all.

Drawing on the ideological explanation of the survival of eugenics, it is suggested here that Swedish social democracy and Zionist Labor movements hardly foster elimination of the "others." The feeble and the poor were part of society as well. However, since they constituted the social margins, the basic idea was to encourage the healthy and strong parts of society to reproduce in order to set the underlying productive basis of society. In Denmark and Finland we find the same mechanism.

However, there are critics of this approach. Critics such as Per Haave in his study on the history of sterilization in Norway, and Mattias Tydén, in his study of the history of sterilization in Sweden, suggested that there is no evidence of a relationship between sterilization practice in the Scandinavian countries and Social Democracy and welfare-state policies (Haave 2000, 2001; Tydén 2002).

This approach came in particular as a critic of the journalist Maciej Zaremba, who drew a parallel between the Scandinavian sterilization policies and those of Nazi Germany. The welfare state and its most clear advocates, Social Democrats, had been the driving force behind the policies of eugenic cleansing.[24]

Zaremba correctly suggested that the impressive belief in the progressive potential of science and social planning had made sterilization by force an attractive tool of policy. Hence, programs of sterilization by force were not confined to authoritarian regimes but also found in liberal tradition countries.

Other critics suggest that the welfare state was not related to eugenics, but after World War II was precisely the factor that led to the end of eugenic practices.

Sven Lindqvist, for example, suggests that welfare reforms of the postwar years led to a new approach toward groups of marginalized, impoverished people on whom the sterilizations were perpetrated (Lindqvist August 30, 1997). As noted, politicians and technocrats believed that the later profligate breeding would gradually come to outnumber the "genetically" healthy. However, that situation "disappeared" when school lunches were freely distributed, housing standards were improved, child allowances were instituted, full employment was upheld, and public health care became accessible to everyone.

In slight opposition to Zaremba, we claim that productivist eugenics differ from Nazi racial eugenics. In opposition to Lindqvist, we suggest that a eugenic renascence is taking place nowadays precisely under the crises of the welfare state. We claim that the positive side of eugenics reappeared when the examined welfare "productivist societies" became aware that their welfare systems were paying for *non-integrated* immigrants. Under the pressure of globalization on the one hand, and the desire of these states to preserve a national culture and the productive national stock precisely because productivity depends on them, there is a logical necessity to increase the breeding of the mentioned national stock. This will come as a replacement of "open" immigration. Irving Fisher, the progressive president of the American Eugenic Research Association, claimed: "I should, as an economist, be inclined to the view that unrestricted immigration … is economically advantageous to the country as a whole …" But, Fisher also added that "The core of the problem of immigration is … one of race and eugenics," the problem of the Anglo-Saxon racial stock being overwhelmed by racially inferior "defectives, delinquents and dependents" (Fisher 1921: 226–227).

This type of thinking was prominent during the "progressive era" in the United States; however, it hardly prospered afterwards. In the United States the idea of productionist society is disentangled from the concept of social or cultural homogeneity. Moreover, diversity plays a key role in various accounts of why social welfare is more restrained in the US than it is in other European countries (Alesina *et al.* 2001). Productionism and diversity work hand-in-hand.

This is not the case with the type of productive welfare societies we deal with. No doubt, welfare-productivist people's homes in Europe, despite being liberal and in some cases theoretically multicultural as in Sweden, tend to halt immigration and advance the reproduction of their national stock. In more senses than one this claim is akin to T.H. Marshall's claim that welfare-state expansion was possible because of a sense of common national community, and, concomitantly, the mutual obligation citizens felt with respect to each other (Marshall 1950). However, the welfare state in itself is not enough. It has to be productive, and for that purpose eugenics played a central role.

Short of racism, however, the eugenic productivist view set the course for an exclusionist version of inclusion. As explained, that means "all" can be included into the collectivity; however, under strict conditions defined by the productive nation.

It is argued that the concern over societal degeneration has not disappeared. Welfare states may encounter a shortage of resources for health and welfare and therefore may become more interested in eugenic policies in order to dictate the national agenda for marginal social groups in the industrial society.

Although this chapter focused on ideology and politics in Sweden, Demark, Finland, and Israel in the 1930s and 1940s, it can provide a substantive ground for viewing the current discussion of new genetic practices as cost-saving public health resources in developed Western democracies, as directly influenced by eugenic ideology. Indeed, developed welfare nations, challenged by economic competition and the pressure of immigration, might aim to reduce the costs of welfare dependency by using genetic reproductive technologies to ensure that as many children as possible are born with genes that enable them to live full and independent lives with no need for welfare dependence (Fitzpatrick 2001; Willis 2005).

Historically laying the attractiveness of eugenics to social democratic ideology and policy and to national welfare states presented in this chapter can offer a new conceptual ground to examine the possible continuity of eugenics in contemporary human genetics and its current use by welfarist states for national goals. However, we should not forget that social democratic and labor movements in Nordic countries as well as in Israel forwarded nationalist goals. In Finland, where the Social Democrats were not in a position to forward a welfare productionist ideology, a nationalist coalition did so. In other words, there were no extreme gaps between a welfare productionist ideology coming from Social Democracy and that coming from nationalism. In both cases, nationalism is not necessarily racist but productionist as we have seen from the Finnish example.

Thus, the bioethical discourse of changing population policies needs to draw on the transition to economic framing of eugenic-based population policy rather than overtly racial attitudes that secured the prominence of eugenics in current politics. We will address Foucault's biopolitical framework, which shows how bioethics engages in the political discourse of the use of new genetic knowledge and practices to support the reemergence of the eugenic rationale in current population policies in other welfarist regimes such as in Germany, France, and the Netherlands.

The type of people's-home countries that are national communitarian states did not endeavor to promote the racial purity of the nation but rather the national productivist purpose. In both Sweden and Israel, until the 1920s racial hygiene underlined the policy efforts and decisions to legitimize and reinforce the social process through which individuals or groups are constructed as being members of a racially homogeneous state. As such, eugenics intersected with the glory of the reified Nordic or Jewish race. However, when such form of collectivist understanding of eugenics became discredited, another form of collectivist type of eugenics replaced it, namely the social-productivist one. Indeed, in both countries, health practitioners and scientists posited a common ideology of productivism and welfare, integrated with a national concern that the population whose members were mentally or physically sick were profoundly damaging to the survival of the nation state. The Zionist solutions luckily did not come close to those of the Swedish sterilization policies, but managing the Jewish mentally ill in Palestine broadly fit into the eugenic framework of preventing the "problem of unproductive people" through immigration policy or eliminating it by means of exclusionary immigration and repatriation. These solutions were complicated by a lack of medical facilities, limited funds, the presence of an imperial power, and the internal political conflict continually brewing in the land.

In Denmark, as we have seen, there was hardly the feeling of defending a Nordic race at any time. Denmark enjoyed a sure and non-disputed national identity, connected to productivity rather than race. This national productivist trend was protected through eugenic measures that were not even recognized as such. In other words, to the Danish, the eugenic debate was not especially relevant because it was totally embedded in their national productivist identity.

In Finland, eugenics appeared in the ideology of its Swedish minority; however, it became an integral part of the national idea. The non-Swedish majority also adopted the eugenic trend, in what could be defined as a Europeanization process. Indeed, rather than an ideological process, eugenics in Finland came with the insertion of Finland into Europe.

The question is why these policies received approval by the general public. Indeed, sterilization laws in Scandinavian countries with a Lutheran state church encountered little public opposition. Social reformers' ideas on national productivism fit Lutheran conservatives and sometimes also Catholics.

Popular concerns about increase in crime and the number of mentally retarded people contributed to a general consensus across the political spectrum on the necessity of eugenic measures. These issues concerned social reformers, who lean to adoption of eugenic measures. In Israel, the dominance of the Labor movement and its view of Jewish survival through the creation of a strong modern society, are central to understanding eugenic policies. In all of these cases we cannot overlook the exclusionist/racist overtone that can be found both in Zionist demographic thinking and in Swedish Social Democratic studies of population, as well as in Danish and Finnish policies.

However, the question is whether eugenic ideology was part of that racist spirit. In some sense it is indisputably a racist trend. However, a second glance allows us to admit that eugenics served as a tool of assimilation, which as we noted can be considered exclusionism from a liberal multicultural perspective. At that time, however, the policies of assimilation represented the opposite of racism. While from a moral philosophical perspective the Swedish and Israelis as well as the Finnish and Danish could be defined as exclusionist, yet we suggest a distinction between racism (racial superiority, racist hatred, and denial of assimilation) and exclusionist national productionism (non-multicultural assimilationism) should be drawn.

Still, the question is not so easy and cannot be wiped out. In 1938, Sweden and Denmark adopted the "J pass," a special passport established by German authorities bearing the J stamp meaning "Jew," which was meant to distinguish German Jews from other Germans with the purpose of barring them from entry into their countries. Danish and Swedish authorities issued notifications to border agencies and embassies that advised against the admission of any person who did not intend to go back to his or her country, who was not economically self-sufficient, or who had no grounds for family reunification. In other words, we are dealing with societies that cannot be freed from yield to Nazi demands. The question is whether these decisions are connected to their eugenic policies. We argue that it is not the case.

In short we are dealing with nation-state countries in which labor social democratic productivist movements, in the main, believe that the only way a welfare state can survive is by reproducing its high quality stock and reducing the expansion of the country's feeble social margins.

3 The bioethical turn to biopolitics

The emergence of bio-nationalism

The historical-institutionalist analysis presented in the previous chapter implies that the adoption of eugenic-based policies should have had a significant impact on the development of population welfare-state policies. The historical look at Swedish, Danish, Finnish, and Zionist labor and social democratic movements during the 1930s and 1940s reveals that despite fundamental differences, elements of eugenic continuity persist by surviving the transition from "racial hygiene" to "national hygiene" politics.

In order to conduct a systematic comparison to address the continuity hypothesis formulated in Chapter 2, we need to analyze the way in which governments have used modern genetic technologies to alleviate their concerns about nationalism and productivity. Thus, using a eugenics-based welfare-productionism category as an intermediate between racism and liberalism, allows us to demarcate a new way of thinking about bioethics that has been developed as a reflective project on moral issues raised by new genetic technologies. Indeed, the analysis presented in this book joins other studies that characterize the gradual transformation in bioethics by an increasing appeal to moral and ethical values and norms that should guide the use of new genetic technologies: What are the ethical groundings of the new genetic technologies' normative discourse? Do the current policies of genetic technologies live up to their moral claims and standards? In reflecting on the moral issues raised by developments in the field of biomedicine and genetics during the 1970s, the bioethical discourse began with criticizing the new genetic technologies to control and assist reproduction (Agar 1995; Caplan *et al.* 1999; Duster 2003; Kerr and Shakespeare 2002; Koch 2003; Raz 2009; Wertz 1998; Wilkinson 2006). However, bioethics could not escape from going beyond the traditional moral philosophy that was concerned with human dignity, end of life, and scarcity of life, as it was associated with the spectrum of political tendencies that aimed at shaping the destiny of people regardless of their private wills.

This book is not concerned with moral issues. This book is inspired by that discussion, but takes a different approach to the issue. While the existing bioethics literature evaluates the normative power of new genetic technologies, this book treats the normative discourse as a device to legitimize the state control over diverse actors both internally (citizens) and externally (states). The nature of this interference differs depending on the governance system of genetic technologies. The policies of genetic technologies are not merely rhetorical exercises: they try to convince the population as a whole that certain policies are good for them, not merely for specific elite. Viewed in this way, they make the marginal population comply rather than rebel. They create normative bonds between the marginal and the rest of the population. These policies are also important for the formation of the national identity. They help states to define their vision of the world and their own role in it. Such policies always have ethical connotations and thereby cannot be seen as products of rational calculations only. But these policies also reflect complex historical and ideological processes that do not correspond neatly with rigid moral positions. The success or failure of genetic technology policies therefore depends on their ability to generate internal and external legitimacy rather than on their ability to meet moral criteria.

For that we suggest using the Foucauldian terminology of "biopower" that responds to population policy problems.[1] In short, in biopower, Foucault referred to certain power/knowledge techniques to rationalize and manage the life of a population. Such techniques emerged in the second half of the eighteenth century, when population gained prominence in various discourses such as demography, statistics, medicine, engineering, and the politics of how to police and control extended groups of human beings in order to empower larger institutional agents (such as the nation state).

The wide discourse of bioethics on the possible continuity of eugenics in contemporary human genetics could be explained by biopower as inspired by different kinds of "bios" with varying deliberative strategies of control and surveillance. People are no longer treated as political or legal subjects but rather as biological or living beings. However, the term bio within the biopolitical underpinning of the new genetics is not to be couched in an explicitly biological sense but rather an understanding of the population at risk. Population discourse in such terms is seen as something that is threatened by internal and external agents, and which can survive and be improved by the elimination of those threats. The kinds of bios that have been discussed in the present book in relation to eugenic-based population policy are racial and national. In this book we suggest viewing the emergence of a bio-nationalism that is not (yet) falling into overt racism but promoting a welfare-productivist democracy. In terms of biopower, that means that when nationalism, or at least the myth of a national community, becomes the engine of productivist socialism, eugenics comes to the aid of the national community in its aim to reproduce itself. The new bioethics needs to engage in the political discourse of reproductive technology—developed people's-home countries in which productivist social democratic and labor movements advocate eugenic exclusionary practices in order to preserve the hegemonic national community.

Productivist eugenics—the new biopower in the service of nationalism

With the current developments in genetic technology, a route into the field of population emerged in the form of productivist eugenics. This book claims that this biopolitical dimension of biopower has the greatest relevance for understanding the ongoing conjunction of genetics and population policy. The attempt to regulate aggregate genetic occurrences at the population level includes the need to respond to issues such as increasing birth rate, declining mortality rate, decreasing morbidity levels, increasing life expectancy, and so on. Foucault's emphasis on the national/everyday functioning of power points to this set of questions—how to strengthen the existing population as a whole in the interests of the nation as an economic, demographic entity—has been increasingly subjected to statistical evaluation from the eighteenth century onwards.

According to Foucault, the issue of population as an object of knowledge and an instrument of the arts of government appeared in the eighteenth century, in the context of population change, the increasing importance of money, and the growth of an agricultural economy. In one early formulation of the question, Foucault claimed that there are two great revolutions in the technology of power: the discovery of discipline and the discovery of regulation. The second of these technologies, biopolitics, emerged in England in the second half of the eighteenth century and targeted not individuals but populations. Foucault understood the concept of population as living beings who are penetrated, compelled, and ruled by a process, by biological laws. "A population has a birth rate, a death rate, an age curve, an age pyramid, a degree of morbidity, a state of health, a population may perish or may, on the contrary, may expand."[2]

Thus, our concern here is to consider Foucault's concept of biopolitics with specific reference to productive nationalism that regards the limits of individual freedom and rights versus social responsibilities and privileges. Population policies of reproduction have occupied an expanding place in European politics and become intrinsic to discussions about national identity on the basis of an overall demographic outcome. It is claimed that nationalism is an intensely biopolitical phenomenon as a process of civilizing people as a nation, a class, a race, and a gender, specifically through control of collectively "individually coded bodies—where they work, how they reproduced, even the language they dream in" (Patel and McMichael 2004: 238). Thus, Foucault's extended account of biopolitics will be used to reveal the not-so-visible techniques of biopolitics that surround reproduction policies in welfarist social democratic countries, through the mechanisms of biopower. Thus, Foucault's critical reflections on biopolitics as an aspect of biopower can clearly be shown to resonate with the welfarist discourse, which simultaneously makes reproduction into an object of study and productive citizens into a subject of power.

Biopolitics—short review of the theoretical background

> [There is a] [l]ongstanding assumption in Western philosophy that there is a fundamental opposition between knowledge and power, that the purity of knowledge can only exist in stark opposition to the machinations of power.
>
> (Farrell 2005: 96)

In this section, we discuss Foucault's concept of power-knowledge to analyze the rules and conditions of formation of the biopolitical population discourse. Foucault looks at the formation of discourse in terms of the "genealogy of problems," which helps to discover "why a certain way of problematizing appears at a given point in time" (Lotringer 1989). Foucault aims to trace the codes of knowledge that constitute a subject (of knowledge and being) by using the methodology of genealogically documenting specific archives that include statements and reported practices, which helps to deduce certain rules of discursive formation and their disciplinary boundaries. Thus Foucault's genealogical study of biopolitical discourse formation is based on the incorporation of institutions and procedures exercising a specific relation of power on the population.

The discoursive formation considers power as a relation, thus producing true rather than just prohibitive discourse. For Foucault, such methodology traces regimes of truth to regimes of power and not to absolute knowledge. The networks of power and knowledge are cited as the formative discourse to regulate the population in order to improve life itself. We increasingly see such internal networks forming to solve contemporary social and political problems facing the public. Foucault suggests that the regulation of population is managed by one of the two major technologies: discipline and biopower. Between these two technologies biopower[3] is the newer technology that is deployed to regulate population, while the disciplinary technology is older and is deployed to regulate individual bodies.

In the following sections, we present the analytical framework in greater detail. We discuss, first, the relation between politics and life that resulted in the production of the discourse of population and then the rules behind the formation of the biopolitical discourse of population (refining its rules of exclusion) and offer the categories of management of population life through biopower technology. Such framework will be particularly based on Foucault's 1976[2003] series of lectures *Society Must Be Defended*, which act as a bridge between two of Foucault's most famous published works: *Discipline and Punish*, which was published in February of 1975 (1975b[1991]), and the first volume of *The History of Sexuality*, which was published in October of 1976 (1989a). In *Society Must Be Defended*, Foucault aimed at connecting his two key technologies of biopolitics: disciplinary power on the one hand, and biopower on the other. In the context of Foucault's *Society Must Be Defended*, especially the eleventh lecture, we can move away from racist to nationalist interest to regulate population, which is put forward in the present book.

Life and politics

Life and death have been discussed since the birth of ancient philosophy. Aristotle (1885, Book I, Part II: 1252) raised the relationship between life and politics to explain the emergence of the *polis* for the sake of life. "The state comes into existence, originating in the bare needs of life, and continuing in existence for the sake of a good life." However, Foucault claims that life becomes challenged within the political scheme through modernity (around the eighteenth century) with the emergence of contractarian theories of government, in which the sovereign gains its right to force its ruling by being capable of protecting the lives and the well-being of people.

> For the first time in history, no doubt, biological existence was reflected in political existence; the fact of living was no longer an inaccessible substrate that only emerged from time to time, amid the randomness of death and its fatality; part of it passed into knowledge's field of control and power's sphere of intervention. Power would no longer be dealing simply with legal subjects over whom the ultimate domination was death, but with living beings, and the mastery it would be able to exercise over them would to be applied at the level of life itself; it was the taking charge of life, more than the threat of death, that gave power its access even to the body … For millennia, man remained what he was for Aristotle: a living animal with the additional capacity for political existence; modern man is an animal whose politics calls his existence as a living being into question.
>
> (Foucault 1989a: 142–143)

The decisive event of modernity, according to Foucault, emphasizes a shift in the structure of power that takes us from the epoch of sovereignty, in which the right of the ruler to take life or let live, to the modern concept of power as a way to enhance, render productive, compose, maximize, and administer life.[4] This shift is conceptualized by Foucault as biopolitics. During the eighteenth century, life formed a crucial contemporary basis for articulating political considerations, occupied with the need to discipline, shape, and regulate bodies in the interest of a "power" in the context of growing capitalist demands for healthy and productive workers.

Lawmakers and other professionals began to develop medical and statistical norms to control life itself by minimizing risk and maximizing life in terms of hygiene, health, and other regulatory controls over the population. Consequently, sovereigns and states developed to become more general managers of life, focusing on the management and monitoring of their populations.

While biopolitics relates to a specific form of exercising political power over life, the term has also become prominent in the context of a wide variety of modernist discourses, including nationalist concerns. Transformation of the national body into an object of political worship gave rise to biopolitics as the symbol of modern theories of national identity—the fusion between the need for biological identification and the quest for nationalism.

Foucault locates the moment at which the discourse of life constituted the national collectivity as a statist totality (Foucault 1976[2003]: 233):

> The bourgeoisie of Third Estate thus becomes the people, and thus becomes the nation. It has the might of the universal. And the present moment—the moment when Augustine Thierry is writing—is precisely the moment when dualities, nation, and even classes ceased to exist.
>
> (Foucault 1976[2003]: 236)

The biological articulation of human existence diffused into the domain of political power. The nation was portrayed as a living organism, functioning according to biological laws and embodying great physical qualities, symbols of innate virtues transmitted from generation to generation (Anderson 1986; Cadman 2009; Turda 2005). The relationship between nation and state outlines allied forms of political citizenship known as "biosociality" or "biological citizenship" that were turned into a specific scientific form of knowledge, one based on biology (Rabinow and Rose 2006). Biopolitics thus operated through investigations of biological processes regulating the triadic relationship between individual, nation, and state, in contrast to Rabinow and Rose (2006) who saw the unifying power of biopolitics in the context of the nationalist discourse. Analyses by Giorgio Agamben (1998) and Hannah Arendt (1943) offer to view the power that lies in nationalism as an exclusionary power to decide who is and is not a member of the political community that is instituted by the state. It is seen that Foucault's biopolitics is a "black box" that could be utilized for various interpretations but, for the sake of argument, we will use the Foucauldian framework of biopolitics through its exclusionary technique that controlling life is translated into a war on internal threats to the national collectivity, in the shape of the people. Thus, the key feature of such technique is concerned with the *population*, rather than the individual body. As could be certainly claimed, there is a certain unilateralism in this determination. One could perfectly well argue that Foucault could be used in order to show the resistance to power politics. Indeed for Foucault, wherever there is power, there is resistance. The idea of resistance in Foucault, however, is rooted in a generic "plebeian instinct" rather than being grounded in specific material positions or specific identities. Foucault tends to focus either on the surface manifestations of power relations or on their underlying technologies. He can hardly provide a clear account of the actual significance of discourses or disciplines in some situations rather than others. While we find occasional references to how the exercise of power in different sites comes to be strategically codified or how capitalist relations of production privileges some disciplinary techniques over others, Foucault is interested in the marginal expressions of power in everyday life (Sum 2004). In this sense, as Marsden has noted, Foucault can be more explicit about the *how* of power but is far less informative about the *why* of power and its contribution to particular forms of social domination (Marsden 1999: 149).

In our case studies we deal specifically with the how of power. However, we could assume that eugenics was the policy of democratic governments; therefore, immediately we should be seeing opposition and attempts to denounce those policies by social groups. Obviously our use of these categories is in a way an attempt to open awareness. However, another important point to be considered in our case is that we have stressed that the "exclusionist" concept of productivist citizens includes groups that had been excluded, such as homosexuals, lesbians, single mothers, and gender in general. These groups that have resisted the power politics of "normality" and that were confined to the fringes, are now included in the agenda of "productivist exclusionist" ideology. The "others" (immigrants), who in Europe are generally coming from Muslim countries, are portrayed as widely supportive of a conservative view towards women, homosexuals, and lesbians, precisely the agents of productivist progress.

In this sense the power of resistance to exclusionist progressive productionism becomes very limited. Such an exclusionist approach to traditional universal welfare, which gives priority to and encourages reproduction of those excluded in the past, becomes difficult to oppose. These sectors thus are the harbingers of policies of "inclusion" of the "excluded," which at the same time limits the power of resistance of the ultimate "other" because their resistance to the "new normality" hardly receives support.

Producing the discourse on population

Turning to the concept of discourse in which discourse produces the meanings of objects and practices, then that discourse is said by Foucault (1972b) to be evidence of a "discursive formation." Foucault's analysis of discursive formations is based on archaeological practice that deals with archives of different areas of knowledge constituted by "what experts say when they are speaking as experts" (Foucault 1982: xxiv; 1989b). As Foucault explains:

> By archives, I mean first the mass of things spoken in a culture, presented, valorized, re-used, repeated and transformed.... How does it happen that at a given period something could be said and something else has never been said? It is, in a word, the analysis of the historical conditions that account for what one says or of what one rejects, or of what one transforms in the mass of spoken things. The "archive" appears then as a kind of great practice of discourse, a practice which has its rules, its conditions, its functioning and its effects.
>
> (1989a: 11)

For example, in "We Other Victorians" (in *The History of Sexuality*), Foucault shows how the discourse around sexuality poses questions concerning the idea of sexual repression and the discourse that surrounds it, from the seventeenth century until today. Foucault challenges the historical evolution of sexual repression during the seventeenth century by piling the layers of the rising discourse on sex since the seventeenth century. His analysis of an emerging proliferation of discourses on sex begins with an examination of the widely held belief that in the Victorian era, sexual experience and practice were subjected to a power of repression. For that, Foucault questions:

> Why has sexuality been so widely discussed and what has been said about it? What were the effects of power generated by what was said? What are the links between these discourses, these effects of power, and the pleasures that were invested by them? What knowledge (*savoir*) was formed as a result of this linkage?
>
> (1989a: 11)

These questions bring into focus the conditions and circumstances of formation of a discourse that produces certain subjects of social attention that lead to a particular ordering of knowledge in specific periods. In Foucault's words, he undertakes:

> … an inquiry whose aim is to rediscover on what basis knowledge and theory became possible; within what space of order knowledge was constituted; on what basis of what historical a priori, and in the element of what positivity, ideas could appear, sciences be established, experience be reflected in philosophies,… what I am attempting to bring to light is the epistemological field, the *episteme* in which knowledge, envisaged apart from all criteria having reference to its rational value or to its objective forms, grounds in positivity and thereby manifests a history which is not that of its growing perfection, but rather that of its conditions of possibility.
>
> (1970: xxi–xxii)

Instead of an epistemological inquiry, Foucault's genealogical analysis of discourse aims at exploring how specific forms of knowledge are complicit with particular forms of power and consequently produce certain knowledge or "truths":

> [A] group of statements which provide a language for talking about—a way of representing the knowledge about—a particular topic at a particular historical moment … Discourse is about the production of knowledge through language. But … since all social practices entail meaning, and meanings shape and influence what we do—our conduct—all practices have a discursive aspect.
>
> (1972a: 107–117)

Exploring the relation of power to the discourse of sex, Foucault asserts that the knowledge/power relation does not essentially lead to the repression of sexuality, but rather causes the production of ever-increasing discourses on sexuality (Foucault 1980: 106). Indeed, in the sixteenth and seventeenth centuries, where stricter code governed the sexual vocabulary in order to purify sex, at the same time the discourse on sex became productive through the practice of confession by the authority of the Catholic Church (Foucault 1988). Through the confessional process, truth and sex have integrated and knowledge of the subject has evolved (Smart 1985). Hence, what seems to be a mechanism of restriction, censorship, and general silencing of sexual discourse is in fact a mechanism of production.

Since the mid-eighteenth century, secular power, too, became part of the proliferation of the discourse on sex. Sex entered the public discourse as a target of deliberate strategies of controlling and regulating populations. A whole network of discourses around sex emerged in the fields of medicine, psychiatry, pedagogy, criminal justice, social work, and demography, which led to population research including measuring birth rate, intervening in marital decisions such as the age of marriage, legitimate and illegitimate births, the frequency of sexual relations, and the need and extent of contraceptive practices. Power/knowledge also manifested itself in the nineteenth century when society developed mechanisms for policing the individual's behavior. Growing interest in school children's sexuality arose in the field of education.

To conclude, the originality of Foucault's genealogical method to analyze the formation of the discourse on sexuality highlights the following general assumptions:

- Knowledge and the power exercised within it work together.
- The object of discourse becomes accessible to knowledge only when power relations established it as a possible object.
- Discourse produces and reinforces power; but it also undermines, exposes, and even blocks it.
- Prohibition, restriction, and censorship are also considered possible effects of power that produce "truths."

As the following sections demonstrate, the eugenic-based productivist population policy discourse is a revealing manifestation of this very discourse formation; it forms a crucial contemporary site for biopolitical framing of current population problems, the generation of new strategies for action to solve them along the complex nexus between the production of knowledge over populations' biological existence, the existing social practices designed to regulate it, and the forms of power relations that are inherent in the shape of national productivist exclusionist discourse.

The rules that governs the discourse

Foucault's analysis of discoursive formations cannot be complete by only emphasizing the productive role of power and the "will to knowledge" in the formation of discourse.

The discourse formation's analysis should concern what governs the power/knowledge regime and legitimizes it. Drawing on *The Discourse on Language*,

Foucault is concerned with the 'rules governing discourse, which include certain external and internal delimitations that have a significant role to play in controlling and structuring all that is said. The most important category among external delimitations or the rules of exclusion is prohibition' (Foucault 1972b: 215–238). Foucault states:

> In every society the production of discourse is at once controlled, selected, organised and redistributed according to a certain number of procedures, whose role is to avert its powers and its dangers, to cope with chance events, to evade its ponderous, awesome materiality. In a society such as our own we all know the rules of *exclusion*. The most obvious and familiar of these concerns what is *prohibited*. We know perfectly well that we are not free to say just anything, that we cannot simply speak of anything, when we like or where we like; not just anyone, finally, may speak of just anything.
>
> (1972a: 216)

Thus the regulations of the discourse, which determine who can tell "the truths" and on what "object" and what "ritual with its surrounding circumstances" (Foucault 1972b: 216), reproduce a whole array of exclusionary practices to divide between "us" and "them," namely between normals versus abnormals, natural and unnatural, insiders and outsiders.[5] Such divisive practices are based on the relational-defining character of power that creates power/knowledge relations with the "others": "The subject is either divided inside himself or divided from others. This process objectivizes him. Examples are the mad and the sane, the sick and the healthy, the criminals and the 'good boys'" (Foucault 1982: 208).

For example, within the discourse on sexuality, Foucault shows how during the sixteenth and seventeenth centuries concern was solely with the category of marriage to limit the discourse's boundaries—such as what one could and could not do within and without the framework of marriage, what were the imperatives for a married couple, and similar issues:

> The legitimate couple, with its regular sexuality … tended to function as a norm … Instead, what came under scrutiny was the sexuality of children, mad men and women, and criminals; the sensuality of those who did not like the opposite sex; reveries, obsessions, petty manias, or great transports of rage.
>
> (1989a: 38–39)

The exclusionary discourse on sexuality uses the category of marriage to divide certain kinds of subjects:

> …from the will to knowledge emerged which … sketched out a schema of possible, observable, measurable and classifiable objects; a will to knowledge which imposed upon the knowing subject—in some ways taking precedence over all experience—a certain position, a certain viewpoint, and

> a certain function (look rather than read, verify rather than comment), a will to knowledge which prescribed (and, more generally speaking, all instruments determined) the technological level at which knowledge could be employed in order to be verifiable and useful (navigation, mining, pharmacopoeia).
>
> (Foucault 1972b: 218)

However, later discourses on sexuality turned the "gaze" to produce "truths" on sexuality outside the category of marriage that were part of the psychiatric discourse on sexuality during that time, including "minor perverts whom nineteenth century psychiatrists entomologized by giving them strange baptismal names" (Foucault 1989a: 43). All this was part of the "scientific" discourse on sexuality of those times.

Such exclusionary practices governing the rules of discourse worked at the unconscious or invisible level: "Only one truth appears before our eyes ... we are unaware of the prodigious machinery of the will to truth, with its vocation of exclusion" (Foucault 1972b: 220). Thus, any analysis of discourse that produces "truths" should also examine what is generally excluded from that discourse, leading to the emergence of a productive exclusionary dialogue.[6]

Exclusionary practices in modern states were shaped by the development taking place in the fields of medicine, statistics, biometrics, security, criminal justice, sociology, politics, geography, and architecture. This brought together the new fields of public health and city planning to address the increasing urbanization of the nineteenth century in a field of enquiry that was framed as "moral statistics" (Crampton 2007). As indicated by Foucault: "It was a matter of organizing circulation, eliminating its dangerous elements, making a division between good and bad circulation, and maximizing the good circulation by diminishing the bad" (2007: 18).

The field of population planning was based on excluding criminals and infertile and poorly educated persons from the discourse on normalizing population. Such exclusion used different framing of "life"—for the "normal" population, the role of power/knowledge was to ensure, sustain, and multiply life, while for the excluded deviant population, the role of power/knowledge was to minimize, eliminate, and totally control life. Such an old regime of normalization and discipline has been transformed today to a new regime of production and reproduction as shaped by the development of genetic technology including genetic screening techniques, cloning, assisted reproduction techniques, and genetic therapies.

Next we turn our attention to the dominant mechanism of power as an agent for the transformation of population discourse nowadays into the political realm of practice and management of population: biopower.

From discourse to practice: the biopolitical dimension of biopower

A good starting point for reconstructing the expansion of the biopolitical discourse in the domain of political power is to distinguish the two distinct manifestations of biopower: sovereign power over individual human bodies and biopolitical power over collectives.

The Foucauldian framework of biopolitics emphasized the transformation of power/knowledge over life from a sovereign's power over death to *foster* life or *disallow* it to the point of death. The sovereign's

> highest function was perhaps no longer to kill, but to invest life through and through.... The old power of death that symbolized sovereign power was now carefully supplanted by the administration of bodies and the *calculated management of life*.
>
> (Foucault 1989a: 139–140, emphasis added)

Such a mechanism of power/knowledge is unlike the sovereign's disciplinary control and surveillance of individual bodies, but is exercised on the population level.[7]

> The mortality rate has to be modified or lowered; life expectancy has to be increased; the birth rate has to be stimulated. And most important of all, regulatory mechanisms must be established to establish an equilibrium, maintain an average, establish a sort of homeostasis, and compensate for variations within this general population and its aleatory field. In a word, security mechanisms have to be installed around the random element inherent in a population of living beings so as to optimize a state of life.
>
> (Foucault (1976[2003]): 246)

The power/knowledge network around the discourse of life produces the discourse on population and thus gives rise to the exclusionary technique that was referred to by Foucault as biopower: "numerous and diverse techniques for achieving the subjugation of bodies and the control of populations" (Foucault 1989a: 140). Foucault concedes that biopower contributed to the rise of capitalism by providing methods for integrating large groups of people into the developing economic system, a process that he calls "the adjustment of the accumulation of men to the accumulation of capital" (Foucault 1989a: 141).

The question remains—how is this management of population life actually practiced today?

Foucault's biopolitical dimension of biopower thus encapsulates eugenics' productivist policy framing of the problem at stake at the population level and generating new strategies for action. The analysis of biopower is constituted on Foucault's four dimensions, including the production of statistical knowledge over populations' activities and existence, together with the social practices, forms of subjectivity, and power relations that inhere in exclusionist discourse and relations between them. The aim of such genealogical analysis of population policy is to describe the conditions of emergence of the biopolitical dimension of population policy and its connections with wider social practices (Foucault 1988: 160). Thus, analysis of power relations associated with population policy can yield insights as to how assisted reproduction technologies came to be accepted as "rational" and "beneficial." Eugenic policies are divided between negative eugenics, which focuses on repressive policies against a specific group, and positive eugenics, which is not used against any special group but is conceived to enhance a dominant one. Drawing on Foucault's ideas on biopolitics, we can focus on productivist eugenic policymaking, which consists of positive eugenic population policies to encourage the best educated and higher earning native-born citizens to have more offspring. Productivist (positive) eugenic strategies will include different forms of practices in which people are given possibilities for choice under the guise of "equal access" and "individual choice." In the practice of funding assisted reproduction technologies and family allowances, positive eugenics can be understood as offering reproductive rights to all citizens, but it is adopted especially for older women of best eugenic value who should be encouraged to increase the number of children they have in order to advance the welfare of the dominant group. A further strategy for promoting positive eugenics includes adoption of a restrictive immigration policy designed to accept only "productive" immigrants admitted on the basis of their skills (Lynn 2001: 220–224).

In short, while a wide variety of countries dealt with eugenics, in the Nordic countries, eugenics, which was inseparable from the work of Social Democrats in office, was integrated within the development of the welfare state since it was intended to enable improvement in the quality of the population while reducing the costs of the policies for social reform and welfare. The implementation of biopower was essential in this process. By implementing biopower within the national-welfarist context, welfarist countries collected data on the birth rates of their populations, their health and longevity, sanitation, and other conditions of environment. The knowledge gathered in this way reinforced the exercise of power by technocratic practitioners who became increasingly interested in eugenic practices as a way to neutralize the margins of industrial society. In this sense we might argue that the synthesis between biopower, technocratic practices, and eugenics permitted the development of successful democratic welfare states.

Towards "bio-nationalist" population policy—measuring eugenic biopower

This book puts forwards the claim that the purpose of biopower is thus no longer to instill fear and discipline populations but, to the contrary, to increase their size, productivity, and health. Thus, we suggest that biopower has become an important tool for European countries, which have followed a track similar to those in Sweden and Israel by using biopower rationalization of eugenic productivist policies to cope with a demographic crisis. The demographic challenge includes changing family structure such as growing consensual marital relationships, couples producing fewer children, and more children raised in households of only one parent. Moreover, declining birth rates have led to an aging population and more mothers are working for pay, placing pressures on both employers and governments to find solutions to childcare and family leave requirements. These challenges question the basic tenets of the welfare states to cope with reducing income taxes, social spending, and "welfare dependency." Although national response to changing demography varies with the institutional pattern of welfare regimes, national governments use the eugenic productivist approach to harmonize their social programs by placing productivist citizens' fertility and reproduction on their agenda and in their regulations. Such an exclusionist approach to traditional universal welfare values is reflected in providing incentives and funding to encourage the reproduction of native-born or long-time settlers.

Given the broad nature of biopower, it is important to clarify at the outset how we will be using it to both compare various national population policies and to assess their convergence into eugenic-based welfare population policy. Thus the next section provides an analytical framework to organize comparative analysis. Four major dimensions of population policy are identified—focusing on policies targeted to certain subgroups of women and couples including control of immigration, funding family allowances, and ART (assisted reproduction technologies).

The bio-problem: shrinking birth rates

The biopoliticalization of welfare nationalism—the manner in which national actors are trying to manage the productivism of populations—is based on detailed statistical knowledge gathered on population birth rate to identify and forecast the population dynamics. Birth rate provides recognition of the population growth in a given country. Birth rate is an important factor for national policy for advancing state intervention to improve its economic development. For instance, in both Italy and Malaysia, national birth rate has been used to legitimize practices such as financial incentives or provision of support services to new mothers. Conversely, China's one-child policy was effectively implemented due to improved information about and availability of birth control methods.

Birth rate, or crude birth rate, is measured by the ratio of live births per 1000 population annually. This figure is combined with the crude death rate to produce the rate of natural population growth (natural in that it does not take into account net migration). Following contemporary surveys conducted by the Population Reference Bureau (2007), the average birth rate for the whole world is 20.3 per year per 1000 total population, which for a world population of 6.5 billion comes to 134 million babies per year.

One way of getting around the problem of population growth would be to focus on TFR—total fertility rate, which is the variable frequency of childbearing in a given population. The total fertility rate (TFR, sometimes also called the fertility rate, period total fertility rate [PTFR], or total period fertility rate [TPFR]) of a population is the average number of children that would be born to a woman over her lifetime if 1) she were to experience the exact current age-specific fertility rates (ASFRs) through her lifetime, and 2) she were to survive from birth through the end of her reproductive life. It is obtained by summing the single-year age-specific rates at a given time. The advantage of total fertility rate over the crude birth rate as an indicator of population growth is that it is not affected by the age distribution of the population (Public Policy Institute of California 2002). However, the TFR is a synthetic rate that does not take into account the total number of children actually born over a woman's lifetime, and therefore cannot predict the actual completed family size rather than the total cohort fertility rate, but instead is based on the age-specific fertility rates of women in their "child-bearing years," which in conventional international statistical usage is ages 15–44 or 15–49. The TFR is therefore a measure of the fertility of an imaginary woman who passes through her reproductive life subject to all the age-specific fertility rates for ages 15–49 that were recorded for a given population in a given year. The TFR measurement refers to the average number of children a woman would have were she to fast-forward through all her child-bearing years in a single year, under all the age-specific fertility rates for that year. Put differently, this rate is the mean number of children born to women of a given generation throughout all their childbearing years.

An alternative fertility indicator can be marked by the net reproduction rate (NRR), which measures the number of daughters a woman would have if they were subjected to current age-specific fertility and mortality rates in the given year. The use of NRR data is greatly advantageous where the number of male babies born is very high, but it also suffers from lack of actual life expectancy deliberation (Kohler 2000).

Across Europe, birth rates have fluctuated in the past decade. The total fertility rate is around two children per woman in every member nation in the European Union. As a result, European populations are either growing very slowly or beginning to decrease. Indeed, in 2000, European populations had the highest percentage of people age 65 or older—15 percent (Kinsella and Velkoff 2001: 9). Eurostat predicted that for the period 2004–2050 the population of the EU is likely to decrease after 2025 and that by 2050, the population will be 2 percent smaller than it was in 2004.[8] Declining birth rates and the growing proportion of the aging population can affect the workforce structure and consequently can reduce productivity and threaten the solvency of pension and social insurance systems.[9]

Control of immigration

The management of immigration and integration touches upon central political issues within the emergence of modern nation states. A growing number of societies are faced with the new migration, which poses a new and perhaps overwhelming challenge to population policy that varies from country to country (Haub 2007). This phenomenon couples both the consequences of the increasingly international nature of world trade and the consequences of war and natural disasters, which result in growing refugee populations. Thus, the first type of immigration is labor-related and/or entrepreneurial for reproduction of labor power (particularly higher and more specialized forms of knowledge), and the second type refers to political asylum across the borders of an increasingly obsolete inter-state system. Each society approaches and deals with immigration and integration issues based on its own nationality law and/or citizenship law. Citizenship in a nation state confers an inalienable right of residence in that state, but residency of non-citizens is subject to conditions set by immigration law. For example, the complex history of global migrations forced powerful states, such as the United States and Canada, to incorporate both legal and political restrictions on the entry and presence of foreigners, while having legal systems in which work without explicit work permission from the state has fallen through legal "cracks." Within the context of a welfare state, social benefits are generally not contingent on citizens' contributions, but citizenship. Indeed, the welfare state is dependent on a high employment level. For that, the non-Western immigrant social group, in particular, is seen as a costly group, whose possibilities of economic self-sufficiency are limited (Tranaes and Zimmermann 2004). It does, however, continuously affect another central element—the ambition of the state to secure its demographic structure based on a relatively ethnoculturally homogeneous population (Fraser 2001). This happens especially in nation-state countries that, despite being multicultural or becoming multicultural in the last decades, want to preserve a hegemonic national culture that is specifically related to the individual rights of the nationals. While immigrants are considered difficult candidates to assimilate into a rational community of citizens and in most cases become dependent on welfare, the tendency of the majority of the population is to support policies that diminish immigration.

Family planning

Family planning services have been defined as involving "educational, comprehensive medical or social activities which enable individuals, including minors, to determine freely the number and spacing of their children and to select the means by which this may be achieved."[10] In practice, family planning policy operates in distinctive ways: the use of birth control, sexuality education, prevention and management of sexually transmitted infections, preconception counseling and management, and infertility management.

The declining fertility rates in industrialized nations after 1960 paved the way for growing acceptance of the idea of planning and timing childbearing. Traditionally, with high rates of infant mortality, larger families held better chances of survival, gaining more status and wealth over time (Kent and Haub 2005). Thus declining fertility rate, high infant mortality, maternal mortality, and low birth weight strained government health services to initiate family planning policies as an efficient way to use scarce resources to raise overall well-being without dramatic increases in health expenditures and the expansion of bureaucracy (Corrêa and Reichman 1994; Furedy 1997; Hartmann 1987, 1997; Smythe 1994). Within the context of economic development, family planning policies have been criticized by feminists and others of the hegemonic economic development critical approach for using family planning to control women and their bodies and deny reproductive rights and freedom of individual choice (Corrêa and Reichman 1994; Hartmann 1987; Smythe 1994). However, the third United Nations International Conference on Population and Development, which took place in Cairo in 1994, represented a shift away from this traditional economic development approach to fertility control through family planning policies, toward an emphasis on individual and particularly women's rights, sustainable development, women's empowerment, access to reproductive services, and the principal of informed consent in the delivery of reproductive health care services.

Assisted reproductive technology (ART)

As a general term, assisted reproductive technology (ART) refers to treatments for many types of subfertility by using artificial or partially artificial means. ART is used primarily as a treatment for infertility but may also be used by fertile couples for reasons related to genetics (e.g., to preselect for fetuses without genetic abnormalities). The Centers for Disease Control and Prevention (CDC), which has gathered statistical data since 1992 related to ART success rates in the United States (Centers for Disease Control and Prevention 1998), defines ART as including

> … all fertility treatments in which both eggs and sperm are handled. Following this definition, ART procedures involve surgically removing eggs from a woman's ovaries, combining them with sperm in the laboratory, and returning them to the woman's body or donating them to another woman.

However, "They do not include treatments in which only sperm are handled (i.e., intrauterine—or artificial—insemination) or procedures in which a woman takes medicine only to stimulate egg production without the intention of having eggs retrieved" (The Fertility Clinic Success Rate and Certification Act of 1992 (Pub. L. 102–493, 42 U.S.C. 263a-1 et seq.).

The growing use of ART techniques since the 1970s posed a major challenge to population planning, and the regulation of reproduction in general. To date, more than 3 million babies have been born worldwide as a result of IVF, and approximately 350,000 IVF treatment cycles occur in Europe each year (ESHRE 2006), The invention of *in vitro* fertilization technology in 1978 transformed the treatment of infertility and subfertility, and contributed to the research of genetic diagnosis of *in vitro* embryos and, more recently, stem cell research (Jamison 2002). The fact that ART technologies by definition involve the handling of human gametes or embryos has opened new debates regarding the resolution of embryonic stem cell research (ESHRE 2002, 2006). Technological innovations such as insemination (e.g., IUI) or fertilization of the oocytes in the laboratory environment (as in IVF) created new forms of parenthood. For example, surrogacy allows gay couples to become parents, and single women or lesbian couples may become mothers by means of a sperm donor (Bleiklie *et al.* 2004a; Bryld 2001). These developments present the state with complex questions regarding the beginning of life and of deciding who is entitled to be a mother, and who deserve to be parents. Echoing the instances of biopolitics, the statistical data of ART treatments forms a basis for deploying "bio-regulation" in an attempt to increase the birth rate of populations.

New assisted reproduction technologies derive support from the positive eugenic perception of national productivism. ART promotes larger families and single parenthood, and groups that take advantage of them are entitled to generous benefits. Thus, this section explains the lax regulation of ART in the Nordic welfare states by reference to the role of the Social Democratic Party in mobilizing public opinion in favor of extensive national coverage for ART treatments and promoting patients' access to ART. Drawing on the underlying biopower rationales of ART policies, we suggest that to some extent certain national-demographic attributes serve as the basis of entitlement to such benefits in the Nordic welfare states.

Following a detailed statistical monitoring of ART in European populations, the average cost of one ART treatment cycle has been estimated to be approximately €3800. The average cost for the procedure itself is approximately €2800, while the cost of medication accounts for the remaining €1000. A recent ESHRE report (2006) on the economic effects of the use of ART shows that these average costs for ART along with the total number of treatment cycles in a given country, make it possible to give a crude estimate of how much of a nation's GDP is associated with the cost of ART.[11] In contrast to countries such as England and France in which national health insurance covers fertility treatments for married women only (Gross and Brammli-Greenberg 2000: 13) the Swedish and Israeli governments, for example, provide these for all women. It is evident that Denmark and Finland follow this direction since the proportion of the GDP associated with ART treatment is substantially higher in Denmark, Finland, and Slovenia, compared with Italy, Ireland, Switzerland, and the UK.

In some countries, attempts have been made to calculate the cost of an "IVF-baby" (Granberg *et al.* 2005). For example, in Sweden, one IVF-child would, on average, cost €22,000, which is 10 percent of the Swedish estimate of health care support of human life. In order to calculate the effect if all European countries were to increase the number of ART cycles up to the levels present in Denmark in 2002, recently the Rand Corporation issued a report evaluating the potential effect of ART used as part of a population policy (Grant *et al.* 2006). The report demonstrates that despite the fact that the demographic effect of ART is rather small, it is comparable in size to the effect of other family-oriented benefits such as child allowances and maternity leave. The report suggests that increasing the number of ART cycles in the UK to levels comparable to those in Denmark would have the same demographic effects as increasing the general child benefit by 25 percent. The statistical data gathered in the report also highlight that the cost of each additional birth was much lower when investing in ART (approx.

€30,000 per capita) compared with increasing the child benefit (approx. €110,000). So, in population policy evaluation terms, increasing the availability and access to ART in a country may indeed be considerably cost effective.

In Denmark, more than 1500 treatment cycles per million inhabitants are performed annually. Spain, Italy, and Portugal, by contrast, perform fewer than 500 annual ART cycles per million inhabitants.[12] Denmark has the highest number of treatment cycles per capita with nearly 2000 ART cycles per million inhabitants in 2002 in Europe.[13]

This review of the dimensions incorporated in positive eugenic population policy will be used in the subsequent chapter to compare and contrast welfarist nations in an attempt to explore commonalities of the modern eugenic policies they embody and the appropriateness of these dimensions in capturing the biopower idea of national productivist eugenics.

4 The bio-regulation return of eugenics for a new century of massive immigration and multiculturalism

The emergence of welfare regimes

In this book we claim that population policies are designed in a biopolitical context where citizens and government agencies share a commitment to welfare-state policy choices, thereby limiting the range of other possible trajectories. Within the context of national welfarist productivism, eugenic ideas take a particular form that concerns subsidizing population and fertility. Focusing on the cases of controlling immigration, funding assisted reproduction technologies, and family planning policy, the present chapter provides an ideational account of this path analysis behind the "locking in" of welfare-state development. Not only does this chapter provide a specific ideational account of convergence policy outcomes across welfare nations, but it also uses the eugenic productivist case as the basis for classification into welfarist regime types.

In the following sections we interpret the state of population policies in contemporary welfare states that are said to be characterized by "national productivist" eugenic-based politics involving demographic and expansion pressures in public benefits and services. Thus, any understanding of reproduction policy development should be consistent with the nature of the national-productivist politics of the welfare state. Population policies are designed in a context where citizens and government agencies share a commitment to welfare-state policy choices in public health, aimed at improving national productivism while unlikely to accept any intrusive or comprehensive population policies that counter this eugenic-based objective.[1]

After briefly surveying the classifications of welfare regimes on the basic tenet of social pending mechanisms of welfare states, we examine the current eugenic-productivist-based population practices aimed at raising birth rates, targeting productive subgroups in society including restricting foreign immigration, funding assisted reproduction technologies, and employing expensive family planning programs. We discuss eugenic policies and their use in the Swedish and Israeli contexts, comparing and contrasting the two in an attempt to explore commonalities of modern eugenic policies. The importance of comparing Sweden's and Israel's eugenic policies first stems from the fact that they are the clearest contemporary examples of biopolitical strategies that are gradually diffusing into other European welfarist countries such as Finland, Denmark, and the Netherlands, as well as Germany and France, focusing deliberately on the convergence in ART, immigration policy, and family planning trends.

Thus, in the next chapter we will use the criteria suggested for eugenics population policies resulting from the present comparative analysis of Sweden and Israel.

Typologies of welfare-state regimes

The aim of universalism has historically been to secure the welfare of the country and its population, by securing basic needs, health, and education in the form of a fair and equitable redistribution of resources and opportunities (Esping-Andersen 1990). The moral added value of the universalist principle in welfarist policy lies in advancing social justice and preserving the dignity of those in need as two of its main ambitions.

Historically, welfare-state studies flourished around the 1970s, when many governments advocated generous funding for social programs by expanding entitlement and creating new programs for lone parents, unemployed workers, and persons with disabilities, which increased social expenditures. Studies of the 1970s demonstrated that welfare programs may impose various ordering of universal elements of welfare practice. However, through the 1980s, many politicians and taxpayers penetrated neoliberal ideas and practices into the realm of decision-making by pressing the need to lower taxes, incorporate reductions in public spending, enhance privatization of government services, deregulation, and enforcement of more "personal responsibility" for economic well-being (Kelsey 1995; Larner 1997). In the 1990s, liberal states focused on reducing income taxes, social spending, and "welfare dependency" (Pierson 1994).

Throughout the 1990s, the policy programs and reforms developed from these ideas were called "welfare regimes" (Esping-Andersen 1990; Korpi 1983; Mishra 1984, 1990; Titmuss 1974). Scholars began to construct typologies to create ideal types of welfare regimes on the basic tenet of universally based mechanisms. Although welfare states have been varied by institutional patterns and political parties and interest groups, similar patterns have been found in the underlying assumptions of social programs, based on why some people are in need, how the state should assist them, and how benefits and services are best delivered.

These patterns could be organized in five categories that constitute the universal framework of welfare policy and thus could be used to show the degree to which a country's welfare regime is universal, earnings-related, or generous[2] (Esping-Andersen 1990; Kenworthy 1999; Korpi 1983). These include:

- "Solidarity"—to compensate families for the economic costs of child rearing;
- Pronatalism—to encourage a higher birth rate;

- "Social justice"—to redistribute income to low-income families with children;
- To protect the well-being of children; and
- In more recent years, to protect parental choice among family types regardless of whether parents choose to work outside the home or to remain at home to rear children.

Following the multidimensional model of welfare policy, Gösta Esping-Andersen (1990) introduced three ideal types of welfare regimes: liberal, corporatist, and social democratic, based on the priorities among these goals, especially between the second and third goals, which have varied over time and space. For instance, the "liberal" welfare regime has continued to stress pronatalism while the "social democratic" regime has emphasized social justice, and the "corporatist" regime emphasizes the goal of solidarity.

Esping-Andersen's categorization addressed explanation of the different types of welfare-state funding of social programs. Following Esping-Andersen's comparison, the fact that some states developed more generous social programs than others lies in a complex combination of strong trade unions and governments dominated by left-wing political parties, which proved to be effective in developing welfare regimes that promote greater income equality.

Esping-Andersen's categorization was followed by intensive scholarly debate. Some researchers have rejected his categorization, arguing that specific countries have been incorrectly categorized. They presented the cases of Australia and New Zealand to show that these states should not have been placed with other liberal regimes because they did not develop social insurance schemes or rely on cash benefits to the same extent (Castles and Pierson 1995). The feminist scholars criticized the way Esping-Andersen's typology used the same category for countries with quite different "caring regimes" or public provision for the care of young children and the frail elderly (Orloff 1993; Sainsbury 1996). Following such criticism, studies have broadened the scope of categorization to compare welfare regimes by focusing on the interrelated dimensions that incorporate a pronatalism goal. As such, Anttonen and Sipilä (1996) referred to the categorization of welfare regimes of public funding for care by drawing the line between private and public responsibility for the care of very young children and the frail elderly. Such criteria emphasized the difference between corporatist states (such as France and Italy) in contrast to other social democratic countries that provide ample services for young children but not for the frail elderly (Anttonen and Sipilä 1996). Leira (2002) used the category of family structure to highlight the differences between welfare regimes by examining different models of family structures. According to Leira's analysis, the "gender-differentiated family" (with fathers earning money and mothers caring for children) is prevalent in the United Kingdom. The family pattern in which mothers' employment is sequential or secondary to that of the fathers is prevalent in France. In Sweden, the dual-earner care-sharing family is becoming more prevalent. Christopher (2002) compared different types of welfare regimes by focusing on the way the variables of tax and transfer systems and employment support in nine Western nations affect the poverty rates within the group of mothers and the group of mothers relative to other groups. The "friendliest" systems to all mothers are found in Sweden and Finland; those in Finland, the Netherlands, and the United Kingdom are the friendliest to lone mothers. However, employment supports for mothers are the most constructive in Finland, Sweden, and France. Christopher (2002) thus concludes that in these countries, mothers enjoy support in women's employment and caring work. Jeandidier and Albiser (2001) compared the impact of family programs on reducing child poverty in the United States, France, Great Britain, and Luxembourg. Highly universal benefits for families with children are prevalent in France and Luxembourg, including support for large families and lone-parent households. In contrast, social transfers in the United Kingdom and the United States are less generous.

However, it is claimed that these typologies share the tendency to highlight differences between countries rather than investigate the inter-contradictions they have. In other words, when explored at the national contextual level, the principle of universalism, which was already characterized as a unique combination of social justice, solidarity, and pronatalism, is bound by strict eligibility criteria such as citizenship or legal residency (Korpi and Palme 1998: 669). When viewed within a single country, welfare programs are more likely to impose various forms of exclusionary practices. Such exclusionary practices result from the unavoidable link between pronatalism and nationalism. Studies have long stressed the connection between pronatalism and nationalism (Yuval-Davis 1997). In nationalist discourses, the nation's strength and authenticity are tied to the biological and cultural reproduction of its people. This makes reproduction central to national projects. Pronatalist politics in "settler societies" such as Israel, Swedish nationalist discourse about "people's home," and the French "republicanism" defined motherhood associated with great support for women and workers. Within the national context, the universal principle of a welfare system may not benefit all social groups equally. We then suggest that for historical and ideological reasons, sometimes even in the name of universalism and equality, the "institutional form" of welfare state guided the welfare policy, leading to an exclusionary productivist type of policy. The extent to which different social groups benefit differently from social transfers, through their working status and citizenship, is determined by ordering the priorities among welfarist goals in which pronatalism has come to the forefront while the social justice concerns have become less visible.

To conclude, nationalism is a crucial factor that determines the degree of productivist eugenics policy convergence. Some welfarist countries are more effective and implement eugenic population policies more quickly than others in the formation of people's interests and identities necessary for their continuation. Variation in policy convergence emphasizes the importance of the institutional form reflected in the response of social democratic or labor movements to changing global economic changes (Baker and Tippin 1999). This is more connected to the social democratic welfarist regime. It is claimed that for more "liberal" welfarist regimes such as Germany and France, it took more time to follow the same track as did the forerunner "social democratic" countries such as Israel, Sweden, and later Denmark, Finland, and the Netherlands. The political foundations for the Nordic welfare states have been closely linked with the Left—especially Social Democracy and left political institutions. As pointed out by Flora and Heidenheimer (1981) and Esping-Andersen (1985), the Social Democratic movement played an essential role in forming the Nordic welfare states. The welfare-state issue has been set on the public agenda in many respects by Social Democratic parties, although other parties—notably Christian Democratic and Centrist parties—adopted many of the pro-welfare issues during the second half of the postwar period (cf. Budge and Robertson 1987). In Denmark, for example, the bourgeois government (1982–1993) did not advocate welfare retrenchment policies; rather it strengthened the welfare state in a Social Democratic policy orientation (Green-Pedersen 1999: 256). In Finland, on the other hand, the Social Democrats in office (1995–2003) responded to the economic crisis by proposing a number of reforms of the public sector as well as direct cutbacks in the social protection schemes (Timonen 2003). Thus, Social Democratic parties face unpopular choices that generally involve withdrawing from accustomed social rights, especially in the Nordic countries. Indeed, the fiscal situation makes them more disposed to pursuing welfare retrenchment in defense of future sustainability and international competitiveness (Schmidt 2000, 2002). Admittedly, the fiscal difficulties faced by these Nordic countries had a significant impact on welfare spending, especially when combined with demographic changes in low birth rate and fertility patterns that are likely to increase the pressure on the Nordic welfare states for productive work through an active labor market (Castles 2004: 4–5; Kautto *et al.* 2001; Korpi 2003: 590; Nygård 2006).

In the following, we briefly review the historical formation of each country's welfare system in adapting to demographic and economic external changes as a background for the biopolitical nature of the ongoing eugenic productivist population policy.

Welfare states under pressure—the economic and political context

More recent studies indicate that despite the withdrawal of a welfare regime and government from the basic tenets of welfare provisions, the commitment to social spending and social transfers to working "productivist" couples and families was not ended (Jenson and Sineau 2001).

There is a broad consensus today about the welfare state being under pressure. Allegedly, the mature welfare state is now facing a world of permanent austerity that puts restrictions on the nation state's capacity to conduct social policies (Pierson 2001; Taylor-Gooby 2004). As previously shown, during the Swedish recession of 1990–1997, the gross domestic product fell and unemployment grew rapidly, especially among single parents. However, poverty rates remained low due to state transfers that helped to compete with parents' declining earning power (Gustafsson *et al.* 2003). Sweden's conservative government introduced only minor reductions in child benefits and wage replacement for parental leave, unemployment, and sickness benefits. When the economy became more stable, the social democratic government reinforced the parental leave scheme and expanded the childcare system even further.

Economic developments during the 1980s and the 1990s have shaken the structures underpinning the Nordic model of welfare, debated the scope for political action in determining the level of public spending on social welfare, and put existing welfare commitments to a severe test (Scharpf 2000; Taylor-Gooby 2004: 30–31). According to Kautto *et al.* (1999), the Nordic Welfare Model features generous public social policy, basic security, earnings-related benefits, a high degree of social security provision, and a combination of flat-rate and equality in income distribution and gender. Thus, it is not surprising that the general impression among scholars is that the Nordic Welfare Model is facing more serious challenges in current times than in the past and that the political elite are forced to make difficult decisions (e.g., Kautto *et al.* 2001; Kosonen 2001; Nygård 2006; Taylor-Gooby 2001).

The possibility of an evolutionary convergence of welfare regimes, as suggested in this book, can be examined within the context of similar economic and political constraints faced by welfarist nations. Thus, this section provides brief economic and political overviews of country-by-country profiles of demographic and social trends that shaped the development of their welfare systems.

Sweden

Historically, Sweden has been following a distinctive and generous pattern of social policy. Sweden's welfare system developed gradually throughout the twentieth century. The origins of Swedish welfare are much debated. Some claim that its origins are associated with the broadening range of social benefits first initiated by Sweden's Liberal Party government (non-socialist), while others emphasize that the development was led by the Swedish Social Democratic Party and the trade unions in the 1920s.

Sweden's welfare state continued to develop during the 1950s and 1960s when Sweden became the second wealthiest country in the world for a period with the achievement of zero unemployment. During the 1973 oil crisis, Sweden had to struggle immensely to maintain that peak. However, in the early/mid-1990s, Sweden faced a major economic crisis. Its unemployment rate "exploded" in the early 1990s reaching a high of more than 8 percent, four times the rate during the 1980s.

The economic growth became negative and the GDP fell. Consequently, the cost of unemployment benefits rose dramatically and the budget deficit boomed. Moreover, the child poverty rate increased from 6.0 percent in 1990 to 8.5 percent in 1994. Although Sweden was brought out of the 1990s' recession by the Social Democrats, in 2000 there was a minor crisis due to the burst of the dot-com bubble.

In 2005, Sweden spent 29.4 percent of GDP on social expenditures (down from a high of 38.6 percent in 1993 and 1994), but still the highest proportion in the EU and OECD (OECD 2010). More than 4 percent of the GDP was spent on family benefits (cash and services) in 2004, in comparison to about 3.5 percent three years later. The poverty rate of lone mothers increased during the mid-1990s, but declined during the end of the decade. By the end of the decade, the child poverty rate had declined to 5.3 percent and stood at 4.2 percent in 2000.

Israel

The Israeli welfare system was first established by the Ministry of Social Welfare in June 1948, carrying on the mission of the Social Welfare Department established in 1931 under the Mandate. Following the British Mandate welfare model, Israel placed great emphasis on the provision of social insurance through free market mechanisms, particularly through the widespread existence of occupational pensions for the elderly, with state programs intended to serve as a safety net and supplement (Doron and Kramer 1991). While the Israeli welfare state was clearly modeled on the British Mandate welfare model when it was first established by the Labor movement during the 1950s, during the 1970s with the introduction of more wage-related benefits, expanded coverage of social demands, and better indexing of benefit levels, it became more similar to the social democratic model. State social expenditure reached 20 percent of GDP in 2003 (Kop 2003). However, the generous nature of the Israeli welfare state and its rate of social expenditure did not yield any effectiveness in alleviating the poverty rate and inequality in Israel, which are still higher than those in most other developed welfare states. In 2002, 17.7 percent of the population lived below the poverty level. In Israel, poverty levels in general, and child poverty levels in particular, are higher than in most other welfare states.

Finland

The 1990s' recession in Finland resulted in decreasing the GDP by more than 7 percent in 1991, caused mass unemployment, and led to a high level of social expenditures (Nordlund 2000; Timonen 2003: 4–5).

Finland shares the Nordic pattern: it is a generous welfare state with a strong family policy commitment. The government spends a (relatively) high percentage of GDP on social welfare, it had a high social protection (social welfare) expenditure of 26.1 percent in 2005, and related, relatively high taxation levels (OECD 2010). During the 1930s, Finland enjoyed remarkable economic growth despite some 1990s cutbacks during the economic recession. And, until the severe economic recession of the early 1990s, Finland had committed itself to further improvements in its policies and programs.

After recovering from the economic recession of the 1990s, Finland strengthened its child and family benefits system, but, according to Taskinen (2003), has not reestablished all family benefits and services to their previous levels.

Finland's welfare policy is also much influenced by an increased female labor force participation (estimated at over 72 percent in 2002) in comparison to other pluralistic democracies, much of it (87 percent) full-time. In this context, child and family policy has gained strong political support. A crucial problem for a welfare system promoting such a work/family agenda, this has resulted in a growing need for comprehensive childcare arrangements, making Finland's under-3 policy a particularly outstanding example. Moreover, Finland's lower child poverty level is managed through income supports, childcare options, preventive health services, and a generally high standard of living. In 2005 Finland's child poverty rate had been reduced to 12.3 percent and the general poverty rate reached a low of 11 percent in 2004.

Denmark

Denmark's welfare system was developed and supported by a strong labor union and Social Democratic Party alliance (and sustained under political change). Denmark was not affected by the recession in the 1990s as were Finland or Sweden (Halvorsen and Jensen 2004: 471–473; Scharpf 2000; Torfing 2001: 302–305).[3]

Denmark had already faced economic insecurity in the mid-1980s, which led the Danish government to introduce a number of welfare-restructuring policies that put constraints on welfare spending and increased the welfare system's resilience to subsequent economic fluctuations (Marklund and Nordlund 1999). Thus, Denmark's mid-1990 economic recovery was followed by a restoration of cuts and further positive developments (Kamerman and Kahn 2000).

Danish social welfare is universal-based with generous and comprehensive policies; accordingly, Denmark's social expenditures as a portion of GDP are high (26.9 percent in 2005).[4] Denmark has one of the highest rates of taxation and of social expenditure as a proportion of GDP among the OECD countries, expending 6 percent of GDP for child and family benefits and services. Denmark is also ranked seventh among 19 OECD countries in per capita GDP, but it is third in a ranking of lowest child poverty due to its generous childcare services. It has a lower poverty rate—5.1 percent. Some 13.8 percent of lone-parent Danish families (15.2 percent of all children) are poor by this standard, compared with 3.6 percent of other families (UNICEF 2000).

Germany

The German welfare system was developed through bargaining and cooperation between the government and employers and labor during the Bismarckian age. Such corporatist structure continues to influence the development of the German economy and the system of social protection with some exceptions, as such structure also combines social market economy with social partnership in industrial relations, and the welfare state has held for more than 50 years (Bahle and Rothenbacher 1998: 37). Social benefits in Germany are closely associated with one's position in the labor market and are largely earnings-related; thus those lacking work histories may not be fully covered. Germany remains among the more generous countries in social protections generally, although it seems that it does not have an explicit family policy and there is no one overall child and family policy statement. Germany spent nearly 26.7 percent of its GDP in 2005 on social protection (OECD 2010). Germany spent 10.5 percent of its social expenditures on family and child benefits in 2004, which is above the EU average of 7.8 percent. Spending on cash benefits for families with children is relatively high, 3 percent of GDP in 2003, compared to an OECD average of 2.4 percent. The expenditure on childcare services, however, is relatively low and is largely allocated to preschool for children aged 3–6 (OECD 2005).

The poverty rate overall in Germany is 14.3 percent, with the average in the west of the country at 12.9 percent, 6.6 percent lower than in the east.[5] In child poverty, Germany ranks tenth, using the European relative measure (50 percent median income). The German child poverty rate is 10.7 percent. German lone-parent families had a 38 percent child poverty rate in 2001; about half of all children in lone-parent families were poor in 1994.

France

The welfare system of France was developed at the end of the 1930s, involving an expanded array of child-related cash benefits and services. The cash benefits are the most extensive and generous in the world and the services allocated are considered the most extensive.

Since 1945, when the family allowances were first introduced, the benefits have increased in number, in selectivity (more income-tested), and in coverage (children are now covered until they are 22, regardless of the employment status of their parents); however, coverage of first children under the basic family allowance was not yet included. There is continued debate as to the effectiveness of family benefits and related policies in easing women's burdens in balancing family and employment.

Despite the economic challenges during the 1980s and the 1990s, which resulted in high unemployment rates, financial pressures, and efforts to reduce the budget deficit, France continued its social protection programs to young families, large families, poor families, and working families. French family policy, however, is comprehensive, including services that are voluntary rather than mandated by law, and a free public preschool system that covers all 3–6-year-olds and almost half the 2-year-olds in a very popular program, relatively high coverage and extensive subsidies for infant and toddler care, and an outstanding maternal and child health system. Social expenditures rose throughout the first half of the 1990s, from 27.7 percent of GDP in 1990 to 29.2 percent in 2005, well above the EU average (OECD 2010). Pensions and survivors' benefits accounted for almost half of social expenditures in 2000, health for almost 35 percent, and family and maternity benefits for 10 percent (housing allowances amounted to almost 4 percent of social expenditures). More than half the expenditures for family benefits are for universal benefits.

When the poverty rate increased during the 1980s and 1990s, the French government introduced the first national social assistance program, called the RMI

(Revenue Minimum de Insertion), in 1989. The RMI is a means-tested safety-net program, targeted to assist individuals aged 25 and older, and those under 25 who have a child they are caring for. This has been an effective program and has been replicated in a number of other European countries (OECD 2005).

The general poverty rate in France in 2004 was 6.2 percent. The French child poverty rate in the mid-1990s was 7.9 percent and slightly less (7.5 percent) in 2001. Using a relative definition of poverty (below median family income), France ranked seventh of 25 countries with regard to child poverty and 15th with regard to per capita GDP.

The Netherlands

The Netherlands was a late developer as a welfare state. During the 1960s the Dutch constructed a strong European welfare state with universal social protection and family programs (Kamerman and Kahn 1997). During the 1960s and 1970s, the government became attentive to changes that occurred within the nuclear family structure, while the diversity of family styles and the rise of cohabitation called for a broader acceptance of individualization, diversity, and pluralism. This approach has continued despite the economic fluctuations that began with the oil crisis of the early 1970s and again during the 1990s by presenting progressiveness in policies and programs supportive of diversity and individuality.

The Dutch welfare system is corporatist in structure through bargaining and cooperation between the government and employers and labor. Aiming at achieving consensus among these parties, the Netherlands has social insurance, medical care, education, social assistance, and housing infrastructure, which, in turn, sustain a comfortable standard for most residents. The Dutch government expenditure (as a percent of GDP) on social benefits was 20.9 percent in 2005 (OECD 2010).

The Netherlands' poverty is a relatively low 4.7 percent of families. This gives the nation one of the lowest poverty rates in Europe. However, this picture is changed when viewing the Netherlands' child poverty rate, which is slightly high—9.8 percent in 2005 (UNICEF 2005).

Within this short review of the economic context, the eugenic productivist reasoning can be found relevant to the deployment of political strategies aimed at improving the economic performance of these welfare nationalist countries. The deeper relevance of the economic challenges on welfarist systems marks a significant extension of political domination over population life. In the following section we draw on current developments in Israeli and Swedish biopolitical regulation of eugenic techniques to support our claim that the national "productivist" concept of eugenics is still present, albeit under a different name; reproduction of the fittest is understood as a national good. The deeper significance of the comparison between Sweden's and Israel's eugenic policies stems from the fact that it marks one of the clearest contemporary examples of how their biopolitical strategies are now gradually diffusing into other European welfarist countries. As we show, the productionist sector of society that the state encourages to reproduce is indirectly described by their ethnicity criteria. Thus in the subsequent section we aim to extend the comparative analysis to other welfare regimes and their relative effectiveness in funding assisted reproduction technologies to increase productivism compared to immigration policies, and that the general philosophy behind welfare regimes appears to be enduring.

Current population management in Sweden and Israel

The use of eugenic policies to promote a welfare-productivist society is not dead. In this section, we explore the present use of eugenics to further democratic ideals and argue that eugenics in current health care provision is linked with the notion of welfare-state growth. We discuss eugenic policies and their use in the Swedish and Israeli contexts, comparing and contrasting the two in an attempt to explore commonalities of modern eugenic policies.

Sweden and Israel are widely regarded as possessing the most ART-friendly environments in the world. ART policies are government sponsored, reflecting deeper demographic concern. In Sweden the aim is to preserve a productivist society while halting immigration. Indeed, Swedish political leaders today seek to control immigration despite the fact that immigrants contribute to the productive development of any modern nation. In Israel, the question of how to preserve the Jewish state is both an economic and a demographic concern. In both cases the underlying motivation behind population policy is to increase the number and welfare of the "ethnically right" population while reducing the number of "others." In both cases, the concept of productivity and ethnicity are more entangled today than in previous years. Yet, the employment of ART in each country takes a eugenic form, which can diverge due to cultural values. This section takes a comparative approach to assisted reproduction practices and regulation along with immigration and family allowance regulations in Sweden and Israel, focusing on the imprint of national and economic priorities on each nation's population policies.

Identifying the bio-problem

The broadening of demographic debate in Sweden was accompanied by growing concerns of decreasing fertility rate and birth rate. Demographers have characterized fertility rates in Sweden as a roller coaster (Hoem and Hoem 1996) due to the rise and fall of the birth rate, which most recently has declined. Delayed family-building and a reduced willingness to have children might be an effect of the difficulties in the labor market during the economic crisis of the early 1990s, when unemployment heavily affected those groups that now demonstrate very low fertility rates (Bernhardt 2002; UN/ECE 1999). In 1999, the total fertility rate (TFR) in Sweden had decreased to 1.52, the lowest rate ever (Swedish Institute 2003), although still slightly above the EU average (Letablier and Pennec 2003). Various studies have examined this fertility decline within the broad

Swedish socio-economic and demographic contexts. A decline in fertility has prevailed among all educational groups except the most highly educated women. Young women, in particular those in the 25–29 age group, experienced a marked fall in fertility (Jönsson 2003). According to Andersson (2000), the decline in first birth fertility can be explained by the decline in fertility among women with very low income. In Sweden, low-income women have a much lower propensity for childbearing than those who are well established in the labor market—well-educated young women. In order to control economic immigration and reduce its demographic effects, massive government expenditure on families with children permitted the rise of "authentic" Swedish fertility. The introduction of the "speed premium" on births led to an increasing pace of childbearing after entering into motherhood. The concepts of a productive society, welfare, and people's-home country thus intimately intersect.

Israeli demographic discourse is also shaped by a preoccupation with ethnicity and nationality, albeit from a different perspective. During the 1970s, demographic research raised the concern of how to secure Israel's unique population composition and the sinking birth rate of the majority population—secular Jews. Several studies, most notably those conducted by Dov Friedlander and Calvin Goldscheider (1979), explored pronatalist population policy as a solution. Legal policies were shaped in the 1950s, 1960s, and 1970s to increase Jewish fertility in order to achieve more rapid population growth. Yet the second half of the twentieth century has shown consistently high fertility patterns only among two segments of the Israeli population: Ultra-Orthodox Jews and Israeli-Arabs. Israel's Arab population boasts a greater total fertility rate than the Jewish majority population. Statistics in 2007 record a TFR of 2.73 births per Jewish woman versus 4.50 births per Muslim woman (Central Bureau of Statistics 2007–2008). Figures showed a sharp increase by 2005, when the TFR reached 2.7, the highest TFR rate in any advanced industrial nation. Moreover, the rate of babies born to single Jewish women has risen to 6.6 per 1000, up from 3.6 in the early 1990s. Most of these were mothers aged 35–39. In contrast, while the number of births in the Israeli-Arab sector grew from 36,500 births in 1995 to 40,800 in 2000, it has since remained at that level. In 2004 an immediate decline in the number of Arab pregnancies was reported. This figure might be partly explained by the effects of government policy restricting grants to only the first two children in any family.

Throughout the decades, the Israeli government policies reflected deep concerns over these figures. In April 1966, the governmental Natality Committee submitted a slate of recommendations to the government. The overwhelmingly pronatalist committee "considered means by which having a large number of children could be reconciled with family welfare" (Friedlander and Goldscheider 1979: 130). In 1968, the government charged an administrative unit, known as the Demographic Center, with implementing the committee's proposals. The Center's guidelines were pursued by the Israeli government, which offers a series of welfare incentives in the form of direct cash transfers, tax credits, and welfare programs to Jewish women. In 2001 the Demographic Center, supported by Minister of Labor and Social Affairs Eli Yishai, decided to address the "demographic threat" and consider how to encourage births among recent immigrants and secular Jewish women. Sergio de la Pergolla, one of the directors of the Center, has stated that he works towards "favoring only Jewish natality within the parameters of seemingly universal policies."[6] In 2002, Shlomo Benizri, then-Minister of Labor and Social Affairs, renewed the Center's activity and research and announced that the main goal of the center is to "preserve the Jewish characteristics of Israel" (Sarig 2002). It was clear that the fact that the Center's mandate only relates to Jewish women, while excluding Muslim women, was driving the ongoing biopoliticization of population policy steadily forward.

Controlling immigration

Sweden's immigration policy is generally regarded as overly restrictive by outsiders and organizations such as the Red Cross, Save the Children, and the Church of Sweden.[7] The current restriction seems at odds with Sweden's demographic challenges—an aging population and declining birth rates. However, our discussion of the biopolitical actors, namely the Social Democrats' welfare-productivism mentality, can explain the apparent paradox.

Sweden could choose to accept low-skilled immigrants from non-European countries to supplement its diminishing population. A recent survey carried out by the Swedish Integration Board (2007) noted that during the past five years, Sweden has received a large number of immigrants, about 12.3 percent of its population (OECD 2007a). The most frequent areas of origin are in Southeastern Europe, the Middle East, Africa, and Asia.[8] Yet, immigrants are not well represented in the labor force, and their employment rate in Sweden is currently below the high employment rate for natives. In Sweden, unemployment is 3.2 percent for native Swedes, 5.3 percent for those born in an EU or EEA country, 9.5 percent for those born in Europe, and 14.2 percent for those born outside of Europe (Swedish Migration Board Statistik 2001: 3–4). Even within migrant groups there are great differences between nationalities.[9] This trend can be explained by the way money and political capital are spent on eugenic policies rather than on providing education and training to help marginalized immigrants achieve greater effectiveness and productivity in the labor market. As indicated, the education system in Sweden marginalizes or fails to adequately prepare immigrants for participation in the workforce (Norlin 2003).

Sweden's eugenic-type policies are not only targeted to raise the national stock, they also aim to decrease the number of unproductive foreigners. By 2005, the Red Cross, Save the Children, and the Swedish Church had criticized Swedish immigration authorities for deporting asylum-seeking families whose young children exhibit "pervasive refusal syndrome," a profound and pervasive refusal to eat, drink, talk, walk, or engage in any form of self-care. These organizations viewed the government policy as a brutal violation of the United Nations Convention on the Rights of the Child, which Sweden has ratified.

Since the 1990s, most Social Democratic governments have attempted to frame the immigration issue as the main cause for Sweden's demographic and economic crises, which eventually paved the way to legitimize more restrictive immigration policies that do not encourage intensive immigrant participation in the labor market, particularly among low-skilled non-Western immigrants (Andersen 2004; Blume *et al.* 2007; Fair 2008; Kristensen 2007; Nygård 2006; Valtonen 2001).

The long tradition of "demographic obsession" in Israel has led to a very similar picture to that in Sweden. The Israeli aim is to preserve the national home; however in the case of Israel, immigration of Jews to Israel is blessed, especially if they arrive trained and skilful. The wave of immigration from the Soviet Union in 1989 was expected to impact the demographic balance between the Jewish and Palestinian communities. Over 1,000,000 Soviet Jews have arrived in Israel, increasing the Jewish population in the face of the high fertility rate in the Arab population. Some argue that Jewish immigration gave a boost to Israeli Jewish demographics. Israel awarded some privileges on the basis of demography (nationality) and productivism. Russian Jews receive all the privileges of citizenship upon arrival and are eligible for extra transfers. At first, this was rationalized as compensation for the cost of their immigration and to facilitate closing the gap with citizens. Such rationale was also evident in the Israeli pronatalism to unpartnered women and non-traditional family formation. As might be expected given our analysis, special arrangements were introduced in 1994 to favor lone immigrant parents. In contrast to child allowances, which were universal, a new policy was introduced that had a work requirement. Indeed, the Law to Reduce Poverty and Income Inequality (1994) substantially raised the level of income-maintenance grants to single parents. These transfers have a stronger effect on female-headed families than among couple-headed families. Such benefits were consistent with Israel's demographic policy, which targeted the Jewish immigrant community that shared a high proportion of single-parent families, in comparison to the Arab community. The eugenic rather pronatalist rationale was further strengthened due to the low rate of female labor force participation of Muslim women (in comparison to Jewish women). Thus, the Israeli governmental support demonstrated "productivist eugenics" in order to exclude the Arab population while benefiting another group—Jewish immigrants.

Animated by a desire to protect the national stock, the Israeli government decided to support the Efrat Committee for the Encouragement of Higher Birth Rates. The Efrat Committee's main objective is to prevent Jewish women from having abortions. By portraying abortion as the tipping point in Israel's demographic war, Efrat "...called upon Jewish women to fulfill their national duty and bear more children in order to replace the Jewish children killed by the Nazis" (King 2000: 143). With the backing of the Israeli government, Israel's principal anti-abortion group "...has been allowed to carry out its activities unhindered on public property and inside state institutions" (Portugese 1998: 143). In 2003, Efrat, now called the International Organization for Saving Jewish

Babies, rejected the Arab Hadash party's bill to use government funds to finance all abortions in Israel. Ruth Tidhar, a social worker in the Efrat organization, attacked the bill issued by Knesset member Ahmed Tibi, for it is considered a ploy to fuel the demographic war and encourage Jewish women to abort their babies (Katz 2003).

The extension of Israeli pronatalism was endorsed through additional financial incentives including child allowance programs throughout the second half of the twentieth century (DellaPergola and Even 2001). Transfer payments to households of children under the age of 14 (later raised to age 18) originated in 1959 in the form of an amendment to the National Insurance Law of 1953. Originally, the child allowances program targeted families with four or more children. However, since Arab families are larger than secular Jewish families, the Israeli administration decided to pursue specific reforms to lower the minimum number of children required in a family to receive payments to three and then to two. Such reforms aimed at supporting the group of secular Jews, the most "productive" of the Jews in Israeli society. The child allowance program was based on a point credit system "under which mothers receive a monthly tax-exempt allowance" from the National Insurance Institute, the allowance equaling the number of credit points to which her children are entitled, multiplied by the cash value of a credit point determined by the number of children in the family below the age of 18 (Child Allowance Law of 1959).

State involvement within the field of reproduction was marked by various financial birth incentives to raise the Jewish birth rate, which excluded the Arab community. The 1968 Fund for Encouraging Birth devised by the Israeli Demographic Center offered young couples wishing to purchase a larger apartment the extension of low-interest loans contingent upon the couple's proof of intent to have another child (Kanaaneh 2002: 36). The Veteran's Family Allowance Scheme was established in 1970; its stated goal was to distribute allowances only to families in which at least one member had served in the Israeli Defense Force or another national security service (King 2000). By doing so it excluded Palestinian citizens of Israel, who are not eligible to serve. Families could receive this allowance in addition to all other child allowances. This program faced substantial political opposition, which eventually yielded in the Knesset vote in 1994 to extend the Veteran's Allowance to all citizens, regardless of prior military service. Though it might thus seem that Israel is moving away from pronatalist policies, in March 1983 the Israeli government enacted the Law for Families Blessed with Children, which gave families with at least four children subsidies and grants (Law for Families Blessed with Children 1983). Following Portugese, from a political point of view, the government actually achieved their social welfare agenda while at the same time encouraging couples to expand family size (Portugese 1998: 104). Thus it seems that the Law for Families Blessed with Children has proved to be a mirror of the state's successful reconciliation of social welfare and pronatalism, which we preferred to view as the enhancement of the eugenic-productivist rationale.

Assisted reproduction technologies

This book puts forward the argument that the new assisted reproduction technologies derive support from the eugenic perception of national productivism. In the context of perceived demographic threats, assisted reproduction technologies are promoted as a means favoring large families and single parents, who are entitled to generous benefits. A brief survey of countries that nurture a technology-intensive fertility industry reveals a wide variety of practices. For instance, the bias around the eligibility criteria for ART public coverage pushes towards constraints on ART use. The bias against non-traditional families appears to have an impact in countries such as England and France in which national health insurance covers fertility treatments for married women only (Gross and Brammli-Greenberg 2000: 13), while the Swedish and Israeli governments provide these for all women. In this section we suggest that this tension that ART helps bring into being, can be explained to some extent by certain national-demographic attributes serving as the basis of the ability to access and enjoy the benefits of ART in both Sweden and Israel.

Assisted reproduction technology (ART) covers a variety of techniques where egg and sperm are not united or an embryo is not created through sexual intercourse but rather through medical intervention. The technique of IVF (*in vitro* fertilization), which was introduced during the late 1970s, has transformed the treatment of infertility and subfertility. IVF technique involves bringing together oocytes and sperm cells *in vitro* and fertilization itself occurs spontaneously (ICMART 2009a). The invention of IVF established the ground for developing further techniques included under ART, such as intra-cytoplasmic sperm injection (ICSI). ICSI was developed in the early 1990s. This technique is used when poor semen quality makes traditional IVF inefficient. In ICSI, a single sperm cell is injected into the oocyte with the aid of micromanipulators. ART also covers frozen embryo replacement (FER) and oocyte donation (OD). Intrauterine insemination (IUI) is a relatively simple procedure that involves placing sperm inside a woman's uterus to facilitate fertilization for some causes of infertility. To date, this procedure has largely been unregulated; the exact number of treatment cycles and subsequent pregnancies is therefore unknown (Palermo *et al.* 1992).

According to the European Society of Human Reproduction (ESHRE 2008), in 2005 it was estimated that over 600,000 ART cycles were performed in many developed countries. The latest survey conducted by the International Committee for Monitoring Assisted Reproductive Technology (ICMART 2009b) estimated that approximately one million ART cycles were undertaken worldwide in 2002, an increase of 12 percent compared with 2000. The growing use of ART has created a complex web of regulatory practices. The United States and Israel are regarded as possessing two of the most favorable ART environments in the world. Israel is believed to have the highest rates of IVF intervention in the world as well as the highest per capita consumption rate of infertility therapy, with *in vitro* fertilization (IVF) at its center (Shalev and Gooldin 2006). Israel provides a relatively unregulated legal environment, including unlimited fertility treatments of all types up to the birth of two living children.[10] The eligibility criteria for such treatments include both single and married women. The financial benefit is added to other benefits for lone parents such as tax, mortgage, and benefits to render childrearing on one salary viable. Thus, while single women in the Jewish community are encouraged to take fertility treatments, only married women in the Israeli-Arab community are allowed IVF treatments. Moreover, hospitals and medical centers in Israel are located in large cities, often far from Arab villages. According to a 2000 survey, nearly half the Jewish women reported visiting a gynecologist during the past year, compared with only 27 percent of Arab women (Myers-JDC-Brookdale Institute 2000). The latest survey undertaken by the Myers-JDC-Brookdale Institute found no gynecological services available in a large number of Palestinian villages and towns in Israel, and that 42 percent of Palestinian women are forced to travel to other localities to obtain these services (Myers-JDC-Brookdale Institute 2000). Ultrasound examinations and neonatal screening tests for pregnant women are usually performed in hospitals located in large cities, relatively far from Palestinian locales. The report suggests that the associated impact of education, health insurance coverage, and distance to family clinics is transparent when measuring the number of visits to the clinics (Myers-JDC-Brookdale Institute 2000). There are also cultural and language barriers for Arab women attempting to access new fertility treatments. Another survey showed that over a third of all Israeli women would prefer to be treated by a female gynecologist, but only about half of those are actually treated by a woman. Among those women who prefer to be treated by a female gynecologist, a greater percentage of Arab women was indicated. However, in Israel only 15 percent of the gynecologists are female. Gross and Brammli-Greenberg (2001) suspect that this figure can explain why fewer Arab women use gynecological services. Furthermore, Palestinian women receive less medical guidance than Jewish women, as most of the available literature is written in Hebrew. Exclusion of Arab women from fertility treatments is seen as a eugenic strategy to exclude Arab women from the Jewish nationalist reproduction project.

Israel's ART-friendly policies follow the recent trends in family forms. Since the Soviet immigration of the 1990s and the increase of single-parent families, the ART policy in Israel has become comprehensive to be able to adapt to the myriad novel family forms that ART helped bring into being. (In the Arab community 6 percent of women are single mothers, in the Jewish community it is 10 percent, according to the Central Bureau of Statistics [2010].) Jewish lesbians are encouraged to undergo fertility treatments. Israel has seen a significant rise in the number of single-parent families, from about 4 percent in the 1970s to about 20 percent in 2003 (Central Bureau of Statistics 2004). The women who choose this form of family structure are mostly well educated and of a high socio-economic status (Central Bureau of Statistics 2007–2008). Thus, the extension of state pronatalism to unpartnered women encourages "productive" individuals of the national stock—educated and employed—to reproduce.

Encouraging homosexuals to have children shows how the pronatal agenda of the government adapted to changes in family structure and planning; there are essentially no Arab lesbians in the public sphere. Paradoxically, groups traditionally discriminated against became "national assets" because they belong to the "right" ethnicity. This policy also illustrates the way in which nationalist policy projects construct notions of insider/outsider and patterns of exclusion by its attention to gender and sexuality. Critical feminist scholars have argued that "race" or "ethnicity" and nation naturalized certain women (e.g., single lesbians) as the physical and cultural reproducers of the nation (Yuval-Davis 1997).

The Israeli government support for ART along with the changes in family structure is expressed in assisted reproduction legislation of artificial insemination by a donor (AID) introduced in 1988. Israel is the first country that permitted the use of AID for unmarried women. While a married woman was entitled to receive these treatments without conditions or restrictions, any unmarried woman was required, by Ministry of Health regulations, to undergo evaluations by a psychiatrist and a social worker as a precondition of receiving AID. Following an appeal to the High Court of Justice, these Ministry of Health regulations were removed.[11] In recent years there has been an enormous rise in the number of applications by single women to sperm banks. The average age of women using these services has gradually decreased from above 40 to above 35 (Mei-Ami 2005).

ART treatment and funding is handled very similarly in Sweden. The government support for ART signals the recognition of an ongoing demographic problem resulting from low birth rate. The growing attention to the demographic issue garnered support for reproductive interventions such as ART. This trend is marked by permissive ART legislation in Sweden that spans two decades—permitting *in vitro* fertilization in 1988, embryo research in 1991, egg donation in 2003, and the creation of embryos for research purposes in 2005 (Kulawik 2004: 17–21). Sweden has been the Nordic leader in assisted reproduction technologies. In fact, Sweden was the first country in the world to pass a law regarding artificial insemination (in 1985). The first Nordic "test-tube baby" was born in Gothenburg in 1982.

As a leading country in the field of assisted reproductive technologies, the Swedish regulation of ART is worth examining within the context of demographic structure. The most progressive party in the area of ART regulation, the Social Democratic Party, has been in power during the crucial periods of legislative action on ART and has promoted permissive regulation of ART.[12] In a 1995 report entitled "Assisted Fertilization" (Statens medicinsk-etiska råd 1995), which was submitted to the Ministry of Health and Social Affairs council, it was stated that Swedish women tend to seek IVF treatment with donated eggs abroad, mainly in England and Italy. The report findings showed the necessity to develop nationally consistent legislation. Subsequently, a bill prepared by the Swedish Ministry of Health and Social Affairs (Prop. 2001/2002: 89) proposed that IVF treatment with donated eggs or with donated sperm be allowed. With the exception of the Christian Democratic Party, all parties were in favor of the bill, and it was approved in April 2002.

In June 2005, the Swedish Parliament, the Riksdag, extended the eligibility criteria to access ART treatments to include lesbian couples. The parliament approved a bill (Prop. 2004/05: 137) allowing lesbian couples to have access to assisted reproduction based on the recommendation of the parliamentary Committee on Children in Homosexual Families, appointed in February 1999. This policy shows growing adaptation to the changing family structure and the need to include more native-born groups that were excluded from assisted reproduction coverage (Häyrinen-Alestalo and Kallerud 2004: 15; Kulawik 2004; Rothmayr *et al*. 2004: 251).

Family planning

Israeli and Swedish cultures both laud the archetype of woman as mother and the woman's maternal duty to national survival. Yet there is a productivist tension between encouraging women to enter the workplace but, at a certain point, to leave their place of employment and embrace their role as child-bearer. Thus, family planning policy is linked closely with employment policy.

Parental leave policies are necessary for a country seeking to increase birth rates while encouraging women to work. In Sweden, a shift from the gender division of labor was reflected in the transformation from the breadwinner family model towards the dual-earner family model during the 1970s. Such transformation in the traditional model of family labor intended to enhance women's participation in the labor market, based on the equality between men and women (Jönsson 2003). This trend is also reflected at the international level. In particular, recent surveys show that the earnings of men and women in Sweden are more equal than in other European countries (European Commission 2007a).

Moreover, studies examining the role of women in the labor market and its associated impact on reproduction show that most Swedish women have a job preceding the birth of their first child (Andersson 2000). Indeed, Swedish welfarist policies were adapted to such trends by designing the paid maternity leave benefits to meet three goals: the well-being of children, women's economic independence, and the involvement of men in family life and childcare (Jönsson 2003). The right to maternity leave was introduced in 1939, becoming paid leave in 1955. The current policy dates from 1974, allowing fathers to share parental leave. When the policy was first introduced, parents were entitled to six months leave, but in the 1970s and 1980s there were several extensions of the period during which the entitlement applied. As of the 1990s, it has been extended to 390 days of income-related benefits and an additional 90 days of flat-rate benefit; these final three months are unpaid, but still job-protected. Non-working parents are entitled to a "maternity leave" of a similar duration, paid at a flat rate that is a bit low (approximately €200 per month).

The Swedish government heavily subsidizes paid parental leave for child sickness and for public daycare. In the 1980s further improvements in parental leave insurance were made, increasing the duration and financial compensation and expanding subsidized public childcare (Jönsson 2003); following these reforms, parents can use parental leave whenever they want until the child reaches the age of 8 years. The Swedish family policy reached its peak in 1985, by introducing the so-called "Speed Premium": "If within 30 months another child is born, then parents are entitled to the same income replacement as for the first child, irrespective of the level of income between the two births" (Jönsson 2003). Although parental leave reforms have been ongoing during the 1990s, they form a massive incentive towards family formation (Oláh 2001).

The Israeli government also offers a series of direct cash transfers, tax credits, and welfare programs targeted at encouraging childbirth. This can be demonstrated by the employment legislation. According to the Employment of Women Law of 1954, the law "prohibits a woman from working during the period of her maternity leave in order to safeguard her health, grants her the right to absent herself from work ... and to a period of leave for adopting a child..." (The Employment of Women Law, 5714–1954, sections 6–7). In 1990 the Employment of Women Law added an amendment, allowing employees to receive sick leave for fertility treatments, with a generous allocation of more than $47 million for the treatment of infertility. This applies to married and single women alike (Gross and Brammli-Greenberg 2000). An additional law passed in 1963, the Severance Pay Law, 5723–1963 (Section 11), entitles a woman to compensation within nine months of giving birth for care of the newborn if she must care for a sick child or marries a man who lives at least 40 kilometers from her home. These legislative initiatives mark a clear preference for productive (employed) single mothers whose numbers have risen within the Jewish community in Israel.

Legal and financial support to motivate a pronatalist regime seems a part of universal health care provisions. However, the suggested overview of reproduction and its associated benefits shows that these policies reproduce race inequalities. It is agreed that Arab citizens are generally technically eligible for social benefits regardless of their racial status; however, Israeli-Arab women face a number of difficulties, which affect their access to reproduction health services and benefits. In a recent survey, Arab women admitted that they did not know of women's health policies including sick leave for fertility treatments (Central Bureau of Statistics 2007–2008). This limited access to the pronatalist package of health and financial benefits can also be explained by the spatial concentration of Arab women in the workforce in peripheral areas exemplified by limited economic opportunities and resources and low rates of female labor force participation (following the Central Bureau of Statistics [2007–2008]). While 56 percent of the Jewish women participate in the labor market, only 13 percent of the Arab women do. As a result, Arab females are excluded from receiving parental leave grants. There is no expectation that Arab women would pursue non-traditional family relations and overcome their cultural and religious barriers to participate in the labor market as the Jewish women do. For the most part, they are expected to remain within the perimeters of the Arab village and thus are not eligible for allowances.

Productivism is seen here as a fluidly interpretive structure, centered on understanding that "productivism" is not universal. Arabs' productivism is evaluated by their labor force participation and not their religious occupation. This can be demonstrated by the case of Ultra-Orthodox Jews (Haredim) in Israel. Ultra-Orthodox Jews have high fertility levels but low rates of male labor force participation; most males spend their time engaging in religious study. Yet Ultra-Orthodox full-time studies are recognized as "productive" by policymakers in Israel. In other words, "productivism" is not universal. Arabs' productivism is evaluated by their labor force participation and not their religious occupation. Thus, productivist policies are really eugenics policies that take strong support from the fact that the Haredim are not productive in the usual sense, but still benefit from the Israeli family policies because of their Jewish identity.

The comparative analysis of current practices aimed at raising birth rates, including restricting foreign immigration, using assisted reproduction technologies, and employing family planning programs in both Sweden and Israel shows that these countries' governments created favorable environments for their people to reproduce and remain productive. In both countries, the need to preserve a people's-home nation should be shouldered primarily by the government, despite the fact that these benefits are viewed by many countries, especially welfarist countries, as a luxury expenditure. The tradeoffs in both countries' policies are clear. To encourage large families among the productive class the government should provide generous incentives and coverage for assisted reproduction programs, bound by strict eligibility criteria such as citizenship or legal residency. Politically and economically, in both countries the government chooses to fund expensive reproductive treatments rather than raise the rate of marginalized others to achieve greater performance and productivism in the labor market.

5 European countries and eugenic productivism gospel—potential convergence trends in eugenic productivist population policy

In this chapter, we extend the comparison of population policy to other social democratic countries such as Finland, Denmark, and the Netherlands, as well as Germany and France, focusing deliberately on the convergence in ART, immigration policy, and family planning trends.

Policy convergence suggests that the strong growth of economic and industrial infrastructures produce similar challenges, which produce similar policy measures and institutions, thus leading to convergence of political institutions, policies, and legal structures. According to Kerr, convergence can be defined as "the tendency of societies to grow more alike, to develop similarities in structures, processes and performances" (Kerr 1983: 3). Drawing on studies of policy convergence among advanced industrial states, convergence seems to occur through various processes that come together with the intent to deal with common policy problems (Barrios *et al.* 2003; Guillén 2001; Holzinger *et al.* 2011; Jahn 2006; Swank 2006; Volden 2006). However, review of a variety body of literature reveals some shortcomings of applying convergence argumentation in comparative policy studies. These shortcomings include very limited systematic understanding about the direction of potential convergence trends, imprecise framing of convergence that obscures complex political and ideological factors, and inability to avoid the interactive processes of feedback that shape policy content (Holzinger *et al.* 2011).

Despite these shortcomings, our focus in this chapter is on convergence of policy content and policy instruments. This is addressed more squarely on comparative case study research rather than on that of an aggregate statistical nature, which is more useful in investigating policy outcomes, particularly levels of public expenditure. Thus, the use of convergence in this present chapter implies a pattern of development over time. We are interested in the extent to which national population policies and levels of regulatory standards have become similar over time across European countries. In the view of the theoretical framework presented in this book, we concentrate on the analysis of three mechanisms that are considered from eugenic productivism as the driving forces of expected population policy convergence: ART regulation and coverage, immigration policy, and family planning. Challenged by economic competition and the pressure of immigration, integration policies in the countries under investigation are under constant review (Bergmark and Palme 2003; Gowricharn 2002; King and Rueda 2008; Koopman 2010; Malmberg-Heimonen and Julkunen 2006; Morrissens and Sainsbury 2005). One eugenic mechanism that may reduce the relative costs of welfare dependency (as a percentage of the GDP) by unproductive people such as immigrants with weaker access to the labor market, is by increasing the reproduction of productive natives who have greater potential to raise economically independent and productive children.

Conjectures on expected patterns of population policy convergence are suggested to be influenced by the fiscal difficulties faced by the selected countries, which have a significant impact on welfare spending. When combined with demographic changes due to low birth rates and fertility patterns, these fiscal difficulties are likely to increase the pressure on welfare states to increase work productivity through an active labor market (Castles 2004: 4–5; Kautto *et al.* 2001; Korpi 2003: 590; Nygård 2006). In the following, we present the emergence of the eugenic productivism perspective and how it materialized in population practices convergent in national welfarist states. Our argument that generous and inclusive welfare states are not beneficial for immigrants may seem counter-intuitive. Indeed, recent studies have shown that generous welfare states are perceived by immigrants as attractive destinations, especially by those with lower education and professional skills (Koopman 2010; Malmberg-Heimonen and Julkunen 2006). Yet, this trend is promoted by various mechanisms that may be misleading, for example, immigrants' tendency to compare their level of welfare deprivation relative to their home country rather than to the citizens of their new state. In reality, a closer look at the data shows that inclusive welfare states tend to yield relatively low levels of immigrant integration into the labor market. Thus, integration policies and types of welfare regimes interact in different ways. Precisely because of immigration and the ensuing transformation of societies into multicultural societies, we can perceive in current times a backlash that takes the form of two complementary trends: civic integration and eugenics.

We proceed as follows. In the next section, we begin by discussing potential demographic reasons associated with low fertility rates that affect the degree of population policy (namely, ART, integration, and family planning policies) convergence across countries under study. The countries selected for this comparison are two of the most generous, inclusive "social democratic" welfare states: the Nordic states of Finland and Denmark, and three of the less generous welfare states that are defined as conservative welfare states: Germany, France, and the Netherlands.

The bio-problem: short overview

Finland

Finland's fertility rate is comparatively high for Europe, 1.85 children per woman (Figure 5.1) in 2008, the same as it was in 1994. Finland's fertility level can be partly explained by policy measures that intend to reconcile the impact of labor-market attachment on entry into motherhood. Finland has one of the highest rates of women's employment in the OECD, over 72 percent (2000), and most Finnish women work full-time. Over 80 percent of couples with children are dual-earners; 50 percent of single mothers are employed and over 90 percent of single fathers also are employed. Half of all children have two working parents, but childless women have lower employment rates than those with children (Statistics Finland 2009).

The age at childbirth rose by one-tenth to 30.1 years (Statistics Finland 2009), when most women are working full-time. The fertility of those under 25 has fallen, whereas that of the over-30s has risen. The fertility rate of women between 25 and 29 has declined in the past few years. The marriage rate (defined as the number of marriages per 1000 adult women of marriageable age) reached a peak after World War II. During the late 1990s marriage rates slightly decreased and divorce rates increased, which resulted in a decrease in the number of wedded couples (Kartovaara and Sauli 2001; Statistics Finland 1997: 11). Despite the declining marital stability and delayed entry into parenthood, Finland has maintained a high level of fertility. The most obvious explanation for Finland's high fertility lies in its particularly active family policy with support for gender equality.

Denmark

The total fertility rate in Denmark in 2008 was 1.89 children born/woman. (Statistics Denmark 2009; Figure 5.1). Denmark has a high female labor force participation rate. In 2000, 71.6 percent of all women 16–66 were in the labor force; only one-fourth worked part-time, below the EU rate. Women make up 46 percent of the labor force, a number exceeded only by Finland. In 1996, 92 percent of mothers and 96 percent of fathers participated in the labor force; 86 percent of women aged 30–50 work full time. The percentage of mothers with children under age 3, in the labor force, is 71.4, similar to the 77.8 percent of those whose youngest child was 3–6. Over 70 percent of lone parents were at work. Denmark's stated policy is to encourage women to enter the labor market, to help women reconcile family and work, and remain in the labor market:

> Danish policy … is aimed towards gender-equitable labour force participation on a full-time and universal basis. To that aim, comprehensive family support is provided, workers can access generous child-related leave programmes, and child care is accessible to almost all children as from the age of 6 months.
>
> (OECD 2002, vol. I: 15)

The age at childbirth of native Danish women is around 29, when most women are working full-time. High divorce and separation rates and socially acceptable cohabitation give Denmark high lone-parent totals; around 19 percent of children were in single-parent families in 2001. As indicated in the case of Finland, marital stability has been declining and childbearing is being postponed. Nevertheless, Denmark offers a generous family policy with support for gender equality.

Germany

As the largest country in the EU, Germany has a total fertility rate of 1.41 children born/woman (2008) (Figure 5.1). Two decades of low total fertility rates have created concern, especially in comparison to non-native Germans; for instance, in 1996, TFR was 1.39 for ethnic Germans and 2.40 (down from 4.4 births per woman in 1970) for ethnic Turks (Mammey and Schwarz 2002). Germany is behind all Nordic countries in percent of women in the labor force; however, this percentage exceeds the EU and OECD averages of 60.2 percent.

In 2005, 67.4 percent of all German women participated in the labor market; half worked of them worked part-time (Erler and Erler 2007). Women employed part-time comprised 81.4 percent of all part-time employees.

Female labor force participation rates fall with the presence of children (Erler and Erler 2007). In comparison to Nordic countries such as Finland and Denmark, Germany's conservative welfare-state regime, which upholds traditional family values and which (until recently) has done little to support the compatibility between work and female employment, could explain why West Germany has been the forerunner in the patterns of very low fertility, high levels of childlessness, and low transition rates to higher-order births. Thus Germany holds one of the highest female part-time employment rates among OECD countries (average 72.1 percent; OECD 2007b). However, there has been a relative increase in employment for married mothers over the past 29 years. In 2005, higher rates (62.7 percent) were more prevalent among women when the age of the youngest child was between 6 and 16, while among women whose youngest child was younger than 2 years, 36.1 percent were in paid work; that percentage rose to 54.8 percent when the age of the youngest child was between 3 and 5 (OECD 2007b).

Despite global changes in the traditional nuclear family, the family structure remains stable in Germany—over 80 percent of German children live with both their natural parents, who are likely to be married to each other (Federal Ministry for Family Affairs 2004). The proportion of children living with married couples has declined in recent decades, as have remarriage rates. According to the 2006 survey, married couples with children still comprised nearly three-quarters of all German families, while the number of single parents and cohabiting couples with children increased by 30 percent over the previous decade (DESTATIS 2007). The survey findings also point out that 42 percent of all families with children in East Germany were single-parent or cohabiting-couple families. Four out of ten children in East Germany had unmarried parents. Germany holds one of the lowest percentages of family households with children in the entire EU (Federal Ministry for Family Affairs 2004). A comparative study of child benefit packages reveals that within the group of German lone mothers in 1999, 39 percent of them were divorced, 22 percent widowed, 13 percent separated, and 27 percent unmarried (Bradshaw and Finch 2002; Federal Ministry for Family Affairs 2004). The age at childbirth is around 28 for native German women, when most women are working part-time (OECD 2008). The case of Germany thus indicates that there is no current delay in entry into parenthood, partly because German family planning does not support women remaining in the labor market.

France

The fertility rate in France has shown consistency with the decline of fertility rates in other European countries (see Figure 5.1). Its total fertility rate declined between the mid-1980s (1.8) and mid-1990s (1.65), when it leveled off and then rose to 1.9 in 2000 and 2001. Since the 1960s, France has started to set an aggressive set of pronatalist policies. It should be emphasized that despite the fact that France now holds the second-highest fertility rate in Europe, behind Ireland with 2.02 children born/woman (INSEE 2008), there is constant debate of a renewed fear of low birth rates in our time (Keilman 2008).

The evidence from France suggests that it should be possible to maintain a reasonably high TFR by coordinated public policy measures in a vast range of policy fields such as gender, economics, employment, and housing, in addition to family policy. Thus women in France, as in the Nordic countries, combine motherhood with participation in the labor market (Hoem 2008). However, in contrast to Nordic countries, French social policies support women staying in the labor force but offer no support for gender equality. The timing of first childbirth is around 29. In the mid-1990s, about 69 percent of married mothers were in the labor force; 82 percent of lone mothers and 57 percent of women with children under the age of six participated in the labor force. Most of these worked full time (Prioux 2005). From a different perspective, 85 percent of women with one child under six were in the labor force in 1997, 82 percent with two children, including one under six, and 56 percent with three or more children (UNICEF 2000).

Divorce rates have steadily increased in France. As stated by Claude Martin, research director at the French National Center for Scientific Research and a member of the European Observatory on Family Matters, “Since 1965, the likelihood of divorce has quadrupled (from 10 to 40 percent in 1997). While 30 years ago just 6 percent of children were born outside marriage, the rate had rocketed to 40 percent by 1997” (Martin 1999). A new legal agreement (the Civil Solidarity Pact) was introduced in 1999 as a measure to “semi-formalize” cohabitation (without legal marriage) to increase couple stability.

In conclusion, France has been able to secure relatively high fertility, despite current tendencies to postpone childbearing and the growing unpopularity of marriage. This trend can be partially explained by active and aggressive pronatalist family planning policies since the 1970s, which were supported by a wide range of political parties. These policies help women have children without being married and continuing to work. The current debate in France over the declining birth rate is definitely not purely a demographic concern, but rather a concern over high unemployment and higher unemployment for women.

The Netherlands

The total fertility rate of the Netherlands is 1.66 children born/woman (2008 est.; Figure 5.1). Netherlands is Europe's part-time work leader. According to Schulze (2003), 63 percent of working women worked part-time, whereas 12 percent of working men were part-timers. The age at childbirth is around 28, when most women are working part-time. Accordingly, the Netherlands has long had low female labor force participation rates but has now grown to above the EU average of 60.2 percent (Dobbelsteen *et al.* 2000). The employment rates of married and lone mothers in the Netherlands are nearly equal. Lone-parent families reached 13 percent of families with children in 2001, which is presumed to be higher compared to other European countries (Bradshaw and Finch 2002; Starobin 2004).

Marital stability has been declining, but there is no significant delay in entry into parenthood. The family policy of part-time work for mothers implemented in the Netherlands is seen as a good solution for the problems of reconciling work and family responsibilities, and thus is considered critical to the continued economic growth of the Dutch economy.

Controlling immigration

Any discussion of population policy would not be complete without some consideration of how labor market and immigration policy fit within the broader context of demographics and productivism. However, current immigration policy has become separated from population policy in general and is no longer

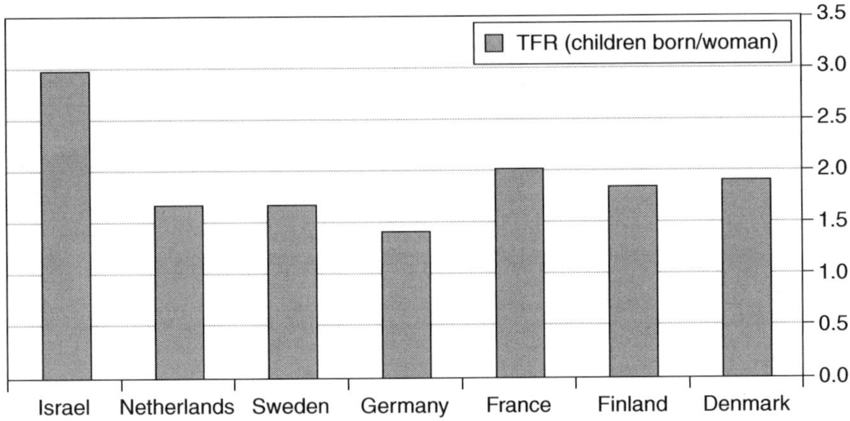

Figure 5.1 Total fertility rate in 2008 (source: OECD [2008]).

regarded as even a partial solution to the problem of low fertility. As stated, recent reforms of extended periods of compensated pre- and post-maternity leave and child/family allowances as part of a national effort to bring about equality between the sexes, reinforce the marginalization of other low-income groups such as ethnic immigrant minorities. The marginal position of immigrants is frequently referred to as a challenge to the welfare state because immigrants are regarded as a costly group whose productivism in the labor market is limited (Table 5.1). In the following we aim to briefly present how each country approaches and deals with immigration and integration issues in its own way, even though activities may be informed by experience in other countries (Malmberg-Heimonen and Julkunen 2006).

Following the typology offered by Castles and Miller (1993), three main models of migration regimes are relevant to our comparison: guest worker, post-colonial, and permanent settler.

The first model is the guest worker migration regime, which is exemplified here by Germany, the Netherlands, and Denmark. In this model, citizenship is an exclusionary measure to restrict immigrants' access to social rights.

In the early 1950s, **Germany** adopted a guest workers' policy. This policy deliberately aimed at attracting foreign short-term contract workers, guest workers, from Turkey and the former Yugoslavia to satisfy the needs of the labor market (Martin 1998). As a consequence of the oil crisis in 1973, immigration policy became more restrictive, and Germany began to pressure guest workers and immigrants to leave. Since the 1980s, however, the number of asylum seekers has increased in Germany (Stalker 2002). Such phenomena resulted in growing social and cultural tensions, as indicated by outbursts of xenophobia and attacks on foreigners in Germany in 1992. As a consequence, Germany introduced tighter restrictions on acceptance to limit unskilled and temporary immigrants, particularly under the conditions of a competitive and unstable global economy. Currently, Germany has an immigrant population of 8.9 percent, most of whom have institutionalized second-class status both in the labor market and in welfare services, which creates dangerous labor market imbalances and puts additional pressure on the state's social security system (OECD 2008).

In **Denmark**, integration and immigration of non-Western immigrants and refugees are gaining central political attention (Kristensen 2007; Østergaard-Nielsen 2003). This group constitutes about 8 percent of the population. The majority, 71 percent, is from non-Western countries (Bauer *et al.* 2004).

Danish immigration policy has also transformed from relatively permissive to restrictive regulation. The critical period that marked this change is during the 1960s and the 1970s. During the economic recession and a lack of labor in the late 1960s, immigrants were encouraged to participate in the Danish labor market on the basis of guest workers. These immigrants came from Turkey, Yugoslavia, and in small numbers from Pakistan. However, market imbalances and an increasing number of immigrants led Denmark to adopt an immigration stop policy in 1973. From the 1970s onwards, the introduction of an immigration stop shows that the government took more selective control of the system of immigration (Blume *et al.* 2007). However, immigration and integration issues have been consistent items on the political agenda and legislative action. In the 1990s Poul Nyrup Rasmussen, then-prime minister of Denmark, leader of the governing Social Democrats from 1992 to 2002, suggested an immigration policy that was bounded by strict eligibility criteria, such as citizenship or legal residency. In 1998, the Act of Integration established the first Danish attempt to create coherent legislation concerning immigrants and refugees (aliens, "foreigners" in the political discourse), covering a comprehensive range of immigration policies including equality (full citizenship), economic independence (labor market participation), and the need to be familiar with "the fundamental norms and values of the Danish society" (cultural knowledge and assimilation).[1] The law emphasized ineligiblity for citizenship as the guiding rule behind the allocation of benefits and allowances to immigrants, which were significantly lower than the common social assistance (*kontanthjælp*) (Lawler 2007). This priority has been endorsed by the two subsequent Liberal-Conservative governments (in power since 2001). Danish immigration and integration policy has long stressed a clear focus on exclusionary labor market participation directly through immigrants' social and demographic attributes.

To be more precise, since the Liberal-Conservative government took office in 2001, citizenship (*medborgerskab*) has been at the center of official discourse, and structures the Ministry of Integration's website. Obtaining citizenship became harder and more demanding, reflecting a steady sequence of legal and administrative requirements. These include an increased residence requirement from seven to nine years (2002), while the government rejected the option that immigrants would hold dual citizenship. Citizenship for second-generation Danish-born residents was conditioned on good marks in Danish (2004) and the tough 2005 Agreement on Citizenship at the behest of the Danish People's Party, which introduced a citizenship knowledge test, a language test, loyalty oath, tougher waiting periods for unemployment, and absence of criminal offences. From 2008, the subjects of the test questions—including conspicuous items on ancient history, national poets, and Christianity—were no longer known and test times were reduced. A minimum language-test mark was also introduced, although its "level 3" requirement already excludes non-academic subjects (Mouritsen 2009: 20).

According to Per Mouritsen, in Denmark the idea of citizenship is not associated with the German ambivalence or self-criticism about membership, or the British regrets about deficient civic traditions (Mouritsen 2009). The idea of cohesion in Denmark is less about community bridging, or national unity, but more about the value of sameness, in itself and as foundational of trust and the welfare state (Mouritsen 2006: 78–79). The values of homogeneity and smallness are indeed akin to the old traditional concepts of egalitarianism and democracy (Østergaard 1992). All of this is often culturalized as an ancient way of life (Mouritsen 2006) that enters into the concept of good citizenship.

Economically, after World War II, the immigration policy of **the Netherlands** shared the same guiding principle as did Germany and Denmark: importing immigrant laborers to boost production. Unions in the Netherlands became biased regarding the willingness of immigrants and guest workers, who would replace native Dutch workers, to work for lower pay. Despite the strict rules of immigration employment and criteria for work permits, there was widespread employment of illegal immigrants in the Netherlands. Illegal immigration was discussed in the Dutch courts, which tended to rule in favor of immigrants, reasoning that the length of immigrants' residency in the country had sufficient basis for citizenship, despite their previous illegal status (Geddes 2003: 105). Even though the Dutch policymakers appear unlikely to choose to confront immigrants due to the state's lack of economic productivity in response to the 1973 oil crisis, the Dutch government ended labor migration and guest worker programs. The restrictive policy towards immigrants included limitation of work permits and the use of penalties for hiring illegal workers (Geddes 2003: 106). In addition, the government tried to limit family migration by using the criterion of minimum household income before one could bring a marriage partner into the country. However, these government regulatory attempts faced difficulties in implementation and enforcement by state authorities (Geddes 2003: 106). Therefore, the Netherlands' immigration shows inner contradictions—on the one hand, through the 1990s, the immigration minority policy was replaced by a more universal integration policy. The integration policy aimed to enhance social networks linking ethnic minorities to the labor market so that minorities would become active citizens. As part of its social inclusion policy, the Dutch government offered courses in Dutch language and culture that were obligatory for the vast numbers of minorities, especially those who were dependent on welfare (Engbersen 2003: 63).

On the other hand, between 1965 and 2004, the 1965 Aliens Act regulated the influx of immigrants through strict rules that prospective residents pass a Dutch language and culture test in their native countries as a condition for admission.

The tightening of asylum policies by the Dutch government targeted illegal-alien employment and increased residency-permit fees by 600 percent (OECD 2004). In addition, the center-right government planned to expel about 26,000 people who had been allowed to stay for years after their asylum applications were not approved. The government policy establishes detention centers for asylum seekers.

The case of the Dutch asylum policy is particularly remarkable given the traditional image of the Netherlands as one of the most tolerant countries in Europe towards immigrants. Indeed, until recently, Dutch citizenship was relatively easy to attain. However, the growing incidence of violence involving foreigners together with a growing concern that their culture is under attack, led Dutch politicians to tighten citizenship and immigration rules. Following a report in December 2003, it is estimated that by 2017 there would be a non-native majority population in Rotterdam. This report led to the continuous racial bias ingrained in growing distrust from the citizens of Rotterdam. Cities' leaders began to review their city building codes, which should address a "less Arabic" design (Ehrlich 2003). This shift in general is expressed by Jan Niessen, the director of the independent Migration Policy Group in Brussels:

> It is as if the Netherlands has realized that they are a multicultural society, and are beginning to say to themselves—"Well, we always said we wanted this, but now we have second thoughts," … The time of formulating nice policies about multiculturalism is over.
>
> (Ehrlich 2003)

The erosion of these demographic concerns that can no longer be ignored is also reflected by Immigration Minister Rita Verdonk. In a speech in parliament discussing whether to force those applying for Dutch citizenship to renounce their original nationality, Verdonk indicated she found that the Islam immigrants should be tagged as a "problem" nation since many innocent people have been proven to have suffered in the name of Islam (Expatica 2004).

The second migration regime is the postcolonial regime, now exemplified by **France**. In this model, immigrants' access to social rights is granted by the state but contested by rules of nationality. France, similarly to Germany, Denmark, and the Netherlands, recruited workers to compensate for local labor shortage. These workers came mainly from the previously French colonies in North Africa. The French colonial immigration policy grants immigrants, at least in formal terms, the rights of citizenship and settlement (Williams 2001). The high turnout of far right nationalist parties in recent elections in France indicates that substantial support may exist for anti-immigrant policies (Kofman 2002).

Since World War II, French immigration policy has favored importing immigrants to satisfy the needs of the French labor market while encouraging the settlement of immigrant families, with an emphasis on long-term settlement and assimilation (Geddes 2003: 53).

Netherlands is not the only state that faced a great influx of non-European immigrants; this trend can be viewed in the case of the decolonization of French territories, which drastically increased the community by the growing number of people from the Maghreb: Morocco and Tunisia, which were French protectorates, as well as immigrants from Algeria, which was considered part of France until its independence in 1962 and therefore had the right to enter and leave France at will. By the late 1960s, a large majority of immigration paperwork and permits was processed inside France (Geddes 2003: 53). The permissive asylum policy of the French government was stated by Jean Marie Jeanneney, the minister of State for Social Affairs, who in 1966 suggested that "Illegal immigration has its uses, for if we rigidly adhere to the regulations and international agreements we would perhaps be short of labor" (Geddes 2003: 53). The fact that the Ministry of State for Social Affairs did not advance comprehensive regulation, combined with a lack of penalties for illegal immigrants, the numbers of migrant workers who moved to France before acquiring the necessary paperwork increased. In 1974 the French government responded to the economic recession by limiting both labor and family immigration. This drawback was followed by an increase in the numbers of illegal immigrants, people who a few years before were allowed to enter France without documents and receive all the necessary legal paperwork from inside the country.

The bias over the criteria needed to distinguish types of asylum seekers has been generated since the late 1960s. It was clear that Europeans, especially Catholics, were considered better able to assimilate smoothly into French society than non-European, often Muslim immigrants. However, the debate over the citizenship eligibility criteria of non-European immigrants appeared to be very controversial, and in a 1968 Economic and Social Council report it was tagged an "inassimilable island" (Geddes 2003: 55).

During the early 1980s the French government emphasized inner contradictions in its immigration policy—on the one hand, she was committed to universal principles such as freedom and civil rights of the immigrant population, on the other hand she pursued exclusionary practices that addressed the rising xenophobic and anti-immigrant public debate that suggested African Muslims were incapable of assimilating to French culture (Wihtol de Wenden 2003: 70). The anti-immigrant sentiment became a popular one in current immigration policy debate. During the major parliamentary elections in 2002, Le Pen's party received nearly 20 percent of the vote in the primary (Murphy 2002; Wihtol de Wenden 2003: 73). Recently, former President Nicolas Sarkozy proposed adding new measures to French immigration law based on how likely an immigrant was to assimilate, which both the Council of Churches of France and representatives from the Muslim community opposed (Van Eeckhout 2006).

The third migration regime is the permanent settler migration regime, exemplified by Sweden and, to some extent, **Finland** (Pitkänen and Kouki 2002). In this model, immigrants' access to social rights is based on a multicultural policy that at least formally grants rights to permanent settlers (Castles and Miller 1993; Esping-Andersen 1990). Finland, in contrast to Sweden and Denmark, has a relatively small immigrant population constituting only 2.962 percent of its population (OECD 2007a). The main ethnic immigrant group (the Ingrians) includes returning migrants, spouses, and humanitarian migrants or refugees. Of the foreign population, the largest groups are from Russia and Estonia (groups that include large numbers of Ingrians), followed by those from Sweden (including returning migrants from the 1960s and 1970s). The main refugee groups, with the exception of groups from the former Yugoslavia, are from non-European cultures hitherto unfamiliar to the majority population, the largest of which are from Somalia (4500 persons), Vietnam (2800), Iraq (2600), and Iran (1700) (Statistics Finland 2000).

The immigrant population in Finland increased rapidly between 1985 and 1994, some years before and after the economic crisis in Finland. Since the 1990s, Finland has pursued an integration policy that seeks to provide immigrants with generous social, political, and cultural rights with the expectation that ethnic minorities will become culturally absorbed into the mainstream Finnish society (Pitkänen and Kouki 2002). In contrast, from 1995 to 2003, the Social Democrats responded to the economic crisis by launching a number of reforms of the public sector as well as direct cutbacks in the social protection schemes, including strict measures to promote integration that would not encourage intensive immigrant participation in the labor market (Timonen 2003).

Recently, incidents of discrimination have been reported by the immigrants in Finland. For example, Valtonen (2001) showed that Finnish employers discriminate against immigrant jobseekers who face barriers when they enquire about a job. Even though the countries included in this study represent various regimes regarding immigration policies, all of these countries exhibit significantly higher unemployment rates among immigrant populations than among non-immigrants. In 1998, the unemployment rate among immigrants was considerably higher in Finland than in the other countries (Graig 2002; OECD 2000).

In conclusion, fears of immigration have not lessened with the expansion of the EU. The debate over Turkey's entrance to the EU exemplifies "old" as "new" concerns shared by EU countries regarding a large influx of "foreigners" if Turkey became part of the EU. This problem is likely to engender a forceful contradiction on the part of welfare states since Turkish foreigners would be EU citizens and thus difficult to keep out, or deny welfare or other citizenship rights. As indicated by past experience upon the entry of Poland to the EU, fears arose concerning the "Polish plumber" (a stereotypical statistic, like the "soccer moms" who became a politically important part of election campaigns).

Even in countries that aimed at extending immigrants' equal membership and the right to full participation in the community through labor market measures such as job training, education, increasing language skills, and support for self-employment, whether such actions actually lead to the real exercise and practice of full social citizenship is another question. Indeed, integration policies directed towards ethnic groups have succeeded only in the creation of low-paying jobs within the public sector, resulting in a class of working poor (Elkeles and Seifert 1996; Gowricharn 2002). As shown in a study by Craig *et al.* (2005), it is suggested that immigrant youngsters are excluded from the benefits of citizenship mainly because of the failure of national integration policies to address discrimination in the labor market.

In Europe, however, the foreign labor force is generally less qualified than the majority population and more often works in blue-collar jobs such as mining, manufacturing, and construction (OECD 2000). As we have seen from the brief overview of immigration policies in the past, during times of economic recessions, unemployment impacts immigrants more than nationals. Immigrants tend to be associated with greater risk of unemployment and work stress because the level of the chief forms of social insurance against sickness, disability, unemployment, and old age, are tied to the level of earnings, while those who cannot participate receive less generous flat-rate compensation (Table 5.1). In practice, chronic unemployment relegates individuals to the marginal area of the social security system.

Viewing these figures from a eugenic productivist approach sheds light on why low-skilled and unemployed immigrants, who are not economically active, cannot help advance these welfare states' economic productivism goals nor help reverse the economic effects that aging and a declining population have effected on these societies. Thus, integration policies in generous and inclusive welfare states could be mediated by welfare states' investment in reproductive technologies as part of

Table 5.1 Unemployment rates of foreign- and native-born populations as a percentage of total labor force

Country	*Native-born*	*Foreign-born*
Denmark	3.4	8.2
Finland	6.7	14.5
France	7.4	13.4
Germany	7.8	14.3
Israel*	8.0	6.0
Netherlands	3.1	7.6
Sweden	5.3	12.1
OECD average	5.7	8.8

Source: OECD (2009a).

Note

* The statistical data for Israel are supplied by and under the responsibility of the relevant Israeli authorities.

the progression towards positive eugenic productivism. ART is but one modern day eugenic tool to achieve this goal.

Assisted reproduction technologies

During the 1960s the issue of assisted reproductive technology (ART) posed a major challenge to the regulation of reproduction. *In vitro* fertilization (IVF) was introduced in the late 1970s, and approximately 350,000 treatment cycles occur in Europe annually; on a world-wide basis, more than 3 million babies have been born to date, marking the beginning of technological development in infertility treatments, and more recently regulating embryonic stem cell research (ESHRE 2002, 2006).

As stated in the first report covering treatments done in 1999, the estimated number of total treatment cycles stood at 169,000 cycles per year (Sunde 2007). In 2002, this number had increased to 324,000 treatment cycles per year (ESHRE 2001, 2006). Although the 19 percent increase is related to the increasing number of clinics reporting, this figure points to a substantial increase in the number of ART treatment cycles performed in European countries (Figure 5.2). ART registry reports estimate that more than 70,000 children are born annually following ART treatment in Europe (Figure 5.2). The impact of these figures indicates that ART should inform regulatory schemes and address its challenges to existing population policy aimed at stabilizing the decline in total fertility rate.

Most comparative research on reproductive policies focuses on three ideal types of policy designs for ART with respect to how strongly the state intervenes into the areas of research in order to explain divergences and convergences among policies (Montpetit *et al.* 2007; Rothmayr *et al.* 2003, 2004; Varone *et al.* 2006).

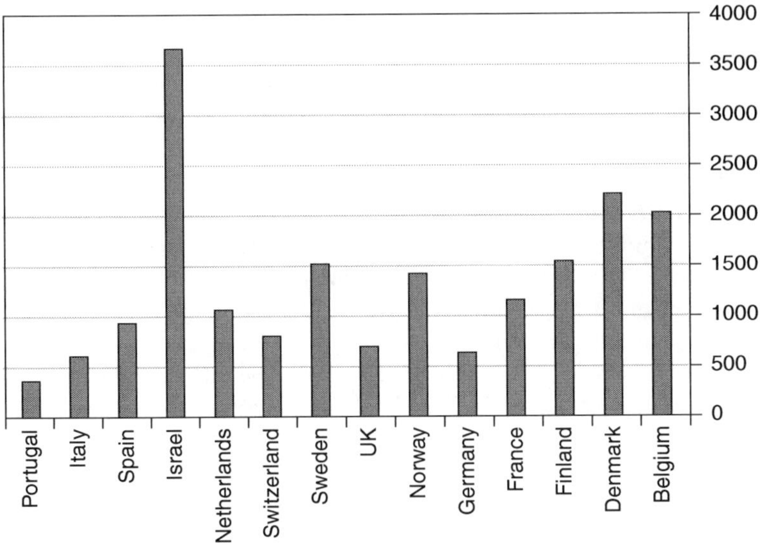

Figure 5.2 Total number of ART* cycles per million capita in 2005 (source: Figures based on http://data.worldbank.org/data-catalog/world-development-indicators?cid=GPD_WDI).

Note
* ART includes here *in vitro* fertilization (IVF), intra-cytoplasmatic sperm injection (ICSI), frozen embryo replacement (FER), and oocyte donations (OD).

The impact of assisted reproductive technologies on underlying political and health care systems, together with socio-cultural, socio-economic, and religious factors across countries varies "from place to place, due to particular local contingencies, both natural and social, political and economic, but also due to different 'ways of life,' different patterns of culture" (Jamison 2002: 100).

The three ideal types of policy design include a permissive ART policy, an intermediate ART policy, and a restrictive ART policy. These ideal types are constructed to identify variations of regulatory scope and content in terms of the policymaking process (Montpetit *et al.* 2007). A policy is considered permissive if the country is pro-technology, for it does not limit research possibilities; the intermediate design would prohibit the creation of embryos through IVF or therapeutic cloning for research purposes but not for reproduction. Such design limits access to IVF by stable unmarried couples. The restrictive design severely limits embryo research by prohibiting all types of invasive research including stem cell derivation, selection of embryos *in vitro*, therapeutic cloning, and gamete donation, and limits access to IVF to married couples with exceptions on a case-by-case basis. The analysis of European countries' ART policymaking in general reveals that Finland is exceptional of all Nordic countries which have implemented policies to regulate the use of assisted reproductive technologies.

In the absence of a comprehensive ART policy design, the practice of assisted reproduction in Finland reflects, by default, the most permissive regime of ART practices in the Nordic region. At the European level, the provision of assisted reproduction in Finland closely resembles the situation in Belgium. The Icelandic government has recently indicated its willingness to move towards a more permissive design (of course, the situation may change drastically if a law is passed eventually). The reverse of permissive design is that of Norway in comparison to the other Nordic countries. At the European level, it can be found in the same restrictive policy design countries such as Germany, Switzerland, and, most recently, Italy. The ART policy design in Iceland and Denmark places those two countries in the intermediate category together with Spain, France, the United Kingdom, and the Netherlands.

Following the present book's outline, by taking the welfare-productivist-eugenic hypothesis, we suggest that we should consider a more eugenic perspective to the analysis of cross-national reproductive policy development. In the following we develop the two dimensions of ART policies—political party role in regulation, and the level of financial coverage and the scope of access to ART, in order to better assess the divergences and convergences among ART policies.

The first analytical dimension concerns the role played by social democratic political parties directly concerned with reproductive regulation. Political parties will initiate a public policy in line with their social-political-ideological preferences. With respect to this dimension, we follow the claim that the left wing is often presented as more permissive when technology is women-friendly. As has been in the case in Sweden and Israel, for example, once the Social Democrats gain electoral power, they favor more permissive regulation of reproduction. According to such criteria, Christian Democrats impose more restrictive policies in order to promote values such as the defense of human life and heterosexual marriage, and delay the decision-making process, if not stop it all together, as in the cases of Switzerland, Belgium, and the Netherlands (Engeli 2009). In Belgium, Schiffino and Varone (2004) show that Social Christians postponed the decision process on ART. Nevertheless, in both Switzerland and Belgium, Christian Democrats were actually unsuccessful in imposing their restrictive regulation in the field of ART.

The second analytical dimension concerns the level of financial coverage for reproductive-related treatment, which can be a private insurance system or a national health insurance system. As mentioned, the average cost of one ART treatment cycle in Europe has been estimated to be approximately €3800. Based on these figures, along with the total number of treatment cycles in a given country, these costs are relatively easily identifiable and quantified to give a crude estimation of how much of the total GDP expenditure represents the ART treatments[2] (Table 5.2).

As we have seen, in Israel and Sweden, ART is part of national population policy. It seems that now Denmark and Finland follow this direction since the proportion of the GDP associated with ART treatment is substantially higher in Denmark, Finland, and Slovenia, compared to Italy, Ireland, Switzerland, and the UK.

Table 5.2 National spending on ART as percent of GDP (crude estimate)

<0.02%	*0.02–0.03%*	*0.031–0.040%*	*0.041–0.05%*	*0.051–0.06%*	*>0.06%*
Latvia	Italy	Norway	Belgium	none	Finland
Russia	Portugal	Spain	Germany		Slovenia
Poland	Ireland	Netherlands	Iceland		Denmark
Bulgaria	Switzerland		Sweden		
	UK		Greece		
			France		
			Hungary		
			Israel		

Source: Sunde (2007: 86).

In some countries, attempts have been made to calculate "the cost of an IVF-baby" (Granberg *et al.* 2005). As mentioned, in Sweden one IVF-child would on average cost €22,000, which is 10 percent of the Swedish estimate of health care support for human life. In order to calculate the effect if all European countries managed to increase the number of ART cycles up to the level held by Denmark in 2002, the Rand Corporation recently issued a report evaluating the potential effect of ART used as part of a population policy (Grant *et al.* 2006). The report shows that although the effect is rather small, it is comparable to the effect of other more financial benefits such as child allowance and maternity leave. In the report, it is suggested that increasing the number of ART cycles in the UK to the level in Denmark would have the same demographic effects as increasing the general child benefit by 25 percent. Such cost–benefit analysis also showed that the cost of each additional birth was much lower when investing in ART (approx. €30,000) compared to increasing the child benefit (approx. €110,000). So, in population policy terms, increasing the availability and access to ART in a country may indeed pay off considerably.

The third dimension considers the patient's access to ART. Access to ART treatment varies widely among European countries. The use of ART creates new forms of parentality—surrogacy allows gay couples to become parents, and single or lesbian women can become mothers through sperm donation. Restricting access to ART treatments to heterosexual couples, in various states such as Italy, France, Germany, and Switzerland, can be explained by the skepticism that exists regarding the impact of ART treatments on the traditional heterosexual family structure, in which women are in charge of nurturing and caring.

These three dimensions are combined in order to compare the regulatory types in the field of reproduction.

Denmark—intermediate regulation of ART

In Denmark, recent data show Danish use of ART to be steadily increasing. In 2002 Denmark provided nearly 2000 ART cycles per million inhabitants (Figure 5.2),[3] exceeding the number of ART cycles performed in Spain, Italy, and Portugal cumulatively, which stands at less than 500 annual ART cycles per million inhabitants (Figure 5.2).[4]

ART treatments are regulated in Denmark by bill L200 on artificial fertilization, which was first introduced to the Folketinget in February 1996. The bill was designed by Poul Nyrup Rasmussen's second government of Social Democrats, Radical Liberals, and Center Democrats. Although the bill includes limits to the entitlements that ART-seeking individuals could claim, as a whole it marks a permissive approach to reproduction. As a February 1996 issue of *The Lancet* reported, the bill "provoked heated debate" in Denmark (Skovmand 1996), and the Christian People's Party tried to postpone its enactment. Thus, the resistance to burgeoning ART use in Danish society led to creation of a new bill, L5, in October 1996 (Damgaard 2004: 123). This bill passed its third reading and was finally adopted in May 1997. Policy deliberations surrounding that bill focused on three issues: 1) the upper age limit for the woman receiving infertility treatment; 2) national coverage of assisted reproduction treatments; and 3) access to fertility treatment of single women and lesbian couples (Albæk 2003: 257; Lüttichau 2004: 83). The Folketinget decided that the "age-limit for women undergoing infertility treatment should be 45 years in both the public and the private sector." It also agreed that "assisted reproduction would be offered at the public hospitals free of charge," although "several politicians from different political corners expressed a wish that this should be a subject for further discussion" (Skovmand 1996).

The issue of equal access to ART treatments by single women and lesbian couples was raised during the first reading of bill L200 by three MPs from the Social Democratic Party who offered to allow access to infertility treatments for women who are married or living with a male partner in a stable relationship comparable to marriage. The amendment was adopted after the first reading of the bill and formulated under Section 3 of the Artificial Fertilization Act: "Medically assisted procreation shall be made available only to women who are married or live with a man in a de facto marital relationship." A study surveying the attitudes among Danish health care professionals regarding the ethical implications of the use of assisted reproductive technologies and legal abortion, found that the Danish medical community insisted on a restrictive approach to ART use with regard to issues such as sex selection, adoption by gays and lesbians, and access to ART by single women and lesbian couples (Fonnest *et al.* 2000). This indicates that the Danish medical community may not have pushed for a more permissive ART policy design, but there has been no organized resistance to burgeoning ART use in Denmark. The bill was not reintroduced to the Folketinget when it reconvened in October 2004.

Governmental support and limited access to single and same-sex headed households need an intermediate ART policy design that combines both permissive and restrictive approaches to ART policy in Denmark. Although ART use is limited to heterosexual couples, loopholes in the law allow single women and lesbian couples to gain access to infertility services via a single clinic in Copenhagen. The practice of ART in Denmark also falls into an intermediate design because of the lack of strict eligibility criteria with respect to age and sexual orientation (Table 5.3).

Finland—permissive regulation of ART

In Finland, as in Denmark, more than 1500 treatment cycles per million inhabitants are performed annually. However, Finland is the only Nordic state that has not yet introduced legislation concerning assisted reproduction. Since the 1980s, several pieces of legislation governing ART have been proposed; however, the sole proposal so far to reach the Finnish Parliament, Eduskunta, resulted in a revolt of sorts that forced the government to withdraw its proposal. The first comprehensive proposal on assisted reproduction in Finland was introduced by a working party (*Oikeusministeriön työryhmä*) appointed by the Ministry of Justice in 1987. The working party deliberations surrounding that proposal focused on two main issues: licensing authority of the National Board of Health to regulate ART treatments and restricting access to infertile married or cohabiting couples or couples whose offspring were likely to inherit a serious disease. The working party suggested including insemination, IVF, embryo transfer, and the use of donated gametes and embryos under ART regulation.

The 1997 proposal resulted in heated public debate surrounding the issue of access by single and lesbian women to assisted reproduction. This issue raised unexpected public interest since the proposal had remained relatively unchanged since its introduction. The decision revolved around what level of access was consistent with the need of the Danish society to face the changes in marital status. Such debate resided with the legal support for non-traditional families provided by other Nordic countries. As both the numbers and visibility of non-traditional families increased during the 1990s, Nordic countries passed liberal legislation allowing couples in same-sex relationships to register their partnerships in a manner consistent with, and with effects similar to, those of marriage. Finland was the in the process of following suit in 2001. The Act on Registered Partnerships of 2001 was under contemplation in Finland at the time, drawing public attention to the legal status and rights of gay and lesbian couples. As in the past, expert opinions were strongly divided on the issues of recipient eligibility, donor anonymity, and now surrogacy.

In Finland, the support of the ruling Social Democratic Party and their collaboration with the medical community to allow Finnish infertility doctors to experiment with a wide range of treatments, lack of public participation in debating the use of ART, no cases of medical misuse, and lack of comprehensive legislation regarding eligibility criteria to access ART treatments,[5] thus entitles single and lesbian couples access to assisted reproduction, leading to permissive regulations in the use of ART. Although the Finnish medical community has been opposed to government regulation of assisted reproduction altogether, the fact that the medical community eventually decided to support a more permissive approach reflects its concern that strict control would deter medical progress in the field, prevent the introduction of new treatment methods, and put an end to

Table 5.3 Current ART policy type

Permissive intermediate	*Intermediate*	*Restrictive*
Finland	Denmark	Germany
Sweden	France	Netherlands
Israel		

some of the treatments that have proven to be effective and beneficial (Table 5.3; Malin and Burrell 2004: 86).

Germany—restrictive regulation of ART

The policy design of ART is governed by the Embryo Protection Act (1990), which prohibits egg donation, embryo donation, preimplantation diagnosis, and transfer of more than three embryos to a woman within one cycle. The law *Embryonenschutzgesetz*, ESchG (Embryo Protection Law, 1990) is regarded as one of the most restrictive ART-regulatory policies currently in force among Western industrialized countries (Kim 2002; Lenoir 1999). The design of ART regulation progressed through four stages, resulting in an elaborated ESchG (Embryo Protection Law). The first stage could be marked by resolutions of the 1985 meeting of the *Deutscher Ärztetag* (German Medical Assembly) that aimed at regulating reproductive techniques following the "Berufsordnung" (professional code). The second stage concerns the resulting report of the Benda Commission (a joint expert group of the Federal Ministry of Justice and the Federal Ministry of Research and Technology) (Benda Report 1985: 1, 49–51). This report articulated the first formal proposition of the Ministry of Justice to elaborate the ESchG. This is particularly noteworthy as it reflected a shift in the German ART policy debate. ART jurisdiction was shared between the Ministry of Research and Technology and the Ministry of Justice. The third stage is marked by the recommendation guidelines report of 1988, convening of the *Bund-Länder Arbeitsgruppe Fortpflanzungsmedizin* (The Federal-State Working Group on Assisted Reproductive Medicine), and finally, in 1988 the ESchG was discussed in the 1988 Kohl Cabinet Report on Human Artificial Fertilization.

The government coalition of Christian Democrats (CDU/CSU) and Liberals (FDP) advocated restrictive regulation of ART. The government coalition followed the constitutional framing of ART as protection of human dignity and human life (Stone Sweet 2000). The Department of Justice took the lead in elaborating the EschG and finally, in the 1990s, adopted and effectively extended the protection to the embryo *in vitro*. This prevailing interpretation of the constitution led to restrictive regulation by the EschG, which actually continued the restrictive approach of the Christian Democrats towards abortion. The position of the Social Democrats was more intermediate towards ART regulation. The Social Democrats prompted restrictions in gender selection of embryos and the use of donor sperm, but did not fully prohibit IVF. The restrictiveness of the

Embryo Protection Law explains the inability of the Social Democratic Party to push less restrictive policy. This could be explained by the competition that emerged between the two Left parties—the Social Democrats and the Greens—that called for a total prohibition of IVF and of any type of embryo research, and the mobilization of feminist and Green interest groups against interventionist reproductive technology (Bleiklie *et al.* 2004b).

In Germany, the eligibility for ART treatment is limited to married couples and heterosexual couples cohabiting for a minimum of two years. The use of donated gametes is restricted to married couples only. Sperm donation is permitted, although payment for such donations is prohibited. Women, however, are restricted from donating their ova (Table 5.3). Furthermore, research related to embryos is narrowly defined and heavily regulated as well (ESchG 1990). The public health care insurance covers 50 percent of the costs for up to three cycles; coverage depends on marital status and age.

France—intermediate regulation of ART

In France ART is regulated by the law on bioethics (Loi No. 2004–800, 2004). The law takes a comprehensive approach to the ethical dimension of biomedicine, addressing issues including, but not limited to, organ donation and stem cell research as well as ART. Initially drawn up in 1994, the law regulating ART (Act 94–654) explicitly defines ART to include embryo transfers or any technology aimed at procreation. However, the act imposes strict conditions on gamete donations and embryo research, along with a call to reduce the coverage of treatments in the national health insurance system but rather encourage greater access by infertile couples. The law of 1994 underwent several revisions (Sandier *et al.* 2004). The Socialist government attempted to render the legislative measures pertaining to embryo research more permissive by separating the issues of fertility and embryo research. However, when the Socialist Party was defeated in the 2002 presidential and legislative elections, the 2004 law reflected the restrictiveness of French ART policy design in prohibiting reproductive and therapeutic cloning and imposed severe penalties on law violators.

The issue of coverage of infertility treatments under the country's health insurance system should be examined within the framework of infertility. In France, infertility is classified as an illness; thus, ART treatment is considered a standard medical intervention, and it is fully covered through the collectively funded health insurance system. French national health insurances only partially reimburse health care expenditure. For most services a co-payment by the patient is required, which is either a direct payment or one covered by voluntary additional health insurance. In 2002 social health insurance constituted 73.3 percent of total health expenditure, with the remainder consisting of voluntary health insurance (13.2 percent), individual payments (9.8 percent), and national taxes (3.7 percent). Infertility treatment and ART provision are, however, exempted from co-payment and reimbursement seems even more generous than for other services. Services in private facilities are covered up to the public amount and patients are expected to pay the difference. Although the additional fee is not reimbursed, many patients can apply for subsidy by a privately funded supplementary insurance fund (*mutuelle*). Doctors are required to inform patients about all such additional costs before any treatment starts.

Currently, there are no restrictions on the number of cycles. However, the recommendation is to limit IVF/ICSI with embryo transfer to four cycles if no pregnancy is achieved (Sorenson 2006). At present, the public health insurance system funds up to six attempts at intrauterine insemination (IUI), provided that the female patient is less than 42 years of age (Agence de la biomédicine 2007).

Coverage is de facto universal and all residents in France have the right to access ART-related services, with the restriction of medically justified age limits and following best practice established by the profession. However, universal access to ART is in contrast to the set-up of the wider welfare and social security system, which is characterized by a strong link between work position (and/or family state) and social entitlements. Residents and citizens who have not contributed to the scheme in turn are only eligible for a means-tested minimum (European Commission 2007b). More modern layers of the welfare system, such as family allowances, are similar to ART support and cover the whole population, irrespective of previously paid contributions and means tests. Nonetheless, given the integration of ART policy into the health care system, France has wider access to ART services than would have been expected from the broader welfare-state context in which it operates. Access to ART treatments is limited to stable heterosexual couples. Same-sex female couples do not qualify for support in ART treatment under this paradigm (Table 5.3).

The Netherlands—restrictive regulation of ART

Dutch policy on ART has emerged since the mid-1980s. In 1983 the first IVF baby was conceived, and later the same year the first pregnancy of a cryopreserved embryo in the world was established. During the 1990s there were over a thousand IVF pregnancies in the Netherlands. Today, IVF is a well-known technique; about one in every 140 babies born in the Netherlands has been outside the womb in its earliest existence, as a result of IVF.

Concerned with the growing IVF treatments in private IVF clinics outside the influence and control of university hospitals, the Christian Political League (the largest political party in the Netherlands, the CDA) rushed to start the legislative process in 1985 to regulate the misuse of genetic material. A provisional act was enacted in 1985 to specify a new set of conditions and controls that confined IVF treatments to hospitals, and then only to a few that were certified by the government. IVF centers that applied for certification needed to meet a number of strict criteria. The government delegated to the Ministry of Health the task of putting together a compulsory licensing system for IVF. The restrictive approach to licensing of IVF practice reduced the number of hospitals that were given a permit to four (apart from the eight universities that report to a different ministry) (Timmermans 2003, 2004). In 1989, a comprehensive "Planning Decree in *in vitro* Fertilization" was enacted. This decree controlled the number of IVF treatments per year (nearly 4500 IVF treatments) allocated among licensed hospitals, guidelines for public coverage for IVF by Health Insurance Council, and a prohibition of commercial trade in eggs. A political window of opportunity opened in 1994, when for the first time since 1945 the Christian Democrats were defeated in the parliamentary elections and a government coalition became secular. The new secular coalition of Social Democrats, Liberals, and Liberal Democrats changed this type of restrictive agenda. Research proposals in this area with respect to the use of embryos and research on human embryonic stem cells were evaluated by the Central Committee for Research Involving Human Subjects (CCMO) according to the requirements of a governmental memorandum of 1995. Indeed, in the early 1990s, the coalition took the lead on regulatory initiatives on embryo research. This regulatory framework provided a platform for a new legal basis for research on human embryonic stem cells with the enactment of Embryo Act in 2002 (Kirejczyk 1999; Timmermans 2004: 170).

The major shift in the government coalition also signified a shift in reproductive policy deliberations, from viewing reproductive medicine as a remedy for infertility to the possibilities for choosing the quality and the characteristics of children born through genetic intervention. Such a permissive approach in this area allowed research on already established stem cell lines, while it became permissible to generate new lines, as long as the embryos used were left over "surplus embryos" resulting from a completed IVF procedure. The Embryo Act prohibits the production of embryos exclusively for research purposes. However, the ban could be lifted five years after coming into force by royal decree. If this should happen, the generation of embryos for research purposes would only be allowed in the future for research which "… would very likely lead to new perspectives in the areas of infertility, artificial insemination techniques, inherited or congenital diseases and/or transplantation medicine, and could only be carried out using embryos as defined in Clause 1." Therapeutic cloning is also currently forbidden in the Netherlands.

Comparatively, the Dutch regulation of ART consists of an intermediate approach. The Social Democrats did use the window of opportunity to change the ART policy of the Christian Democrats to take a more permissive approach. Indeed, the return to power of the Christian Democrats in the fall of 2002 did not lead to withdrawal from the framework put forward by the secular coalition and imposition of their restrictive policy goals (Table 5.3).

Family planning

Since family policies vary widely in their scope and in the number and complexity of their measures, we will focus on the most consensual measures now provided by most countries, including a mandated period of compensated pre- and post-maternity leave (Figure 5.3) and cash allowances for families with children up to the age when they leave school, or beyond that if the child is obtaining job training or higher education.

Maternity, paternity, parental, and family leaves

The parental leave system in **Finland** includes maternity, paternity, and parental allowances. Maternity and parental allowances are paid for parents when the pregnancy has lasted 154 days. These allowances are paid for a period of 263 weekdays. After 105 weekdays, the maternity allowance is replaced by a parental allowance. Parental leave is designed to encourage parents' autonomous choice to split the maternity period. Indeed, the mother should take the first 105 days of the parental leave and the remainder can be used either by the mother or the father. If the father wishes to use the remaining leave, the consent of both parents is required. The amount and scope of compensated pre- and post-maternity leave is calculated on the basis of taxable income, the minimum sum (in 2004) being €11.45 per day (Finnish Ministry of Social Affairs and Health 1999).

The paternity leave can be extended by 1–12 weekdays if the father has used the last 12 weekdays of the parental leave. The extension must be taken in a single period immediately following the parental leave. The idea is to give both parents free choice to continue to work or to stop working to raise their children. Thus, parents can choose between receiving a home care allowance for children under school age, to use municipal day care based on the total gross monthly income of the parents, or the mother or the father can legally take three years of unpaid childcare leave without losing her or his job. The idea is that if both parents are employed, one of them is entitled to shorter working days until the end of the child's second year in comprehensive school. An employer can reject this entitlement only if it would have a significant adverse effect on the source of employment.

In **Germany**, maternity and parental allowances are payable for 14 weeks (6 weeks before childbirth and 8 weeks after childbirth) and continuation of all entitlements coverage with no tax-reduction and no reimbursement payments. The leave is mandatory for 8 weeks following childbirth, with variations for special circumstances. Traditionally, the maternity allowance (*Mutterschaftsgeld*) is financed through social security contributions. Consequently, the maternity allowance paid for a female employee is equal to her average net pay up to €13/day and an additional amount equal to the full wage (if greater) paid by the employer during the maternity leave, while female employees not insured by the federal government receive a one-time maternity allowance amounting to €210 (Federal Ministry for Family Affairs 2007). However, during 2005 the German government introduced major changes in its parental leave benefit, imitating the Swedish scheme. The new benefit structure (*Elterngeld*) is no longer financed through social security contributions but through general taxes. It therefore marks a completely new structure in the German welfare state—guaranteed minimum monthly payment to all parents and a cap on the amount of individual benefit payments of an earnings-related benefit equal to 67 percent of wages (Federal Ministry for Family Affairs 2007). The 2007 reforms introduced to leave employment-related policies reflect the need to support parents. The cash benefits during parental leave are no longer used to compensate poor parents but rather to compensate working parents for a temporary loss of income. Both parents are entitled to receive a maximum of three years of job-protected parental leave after the birth of the child, returning to work after three years at the same conditions but without any wage during the leave. The parental leave is paid for 12 months, with an alternative of receiving decreased payments over a two-year period. Private firms with more than 15 employees are flexible regarding parents because parents have the option of working part-time during their parental leave, with the employer's consent, for a maximum of 30 hours per week (Erler and Erler 2007). The leave and policy regulations thus favor married mothers, and in particular well-paid female employees. However, a recent report conducted by the German Federal Institute for Population Research shows that despite Germany's efforts to increase government subsidies paid for families with children, allowances and benefits for parents who care for newborns, the German birth rate has failed to increase (Bujard *et al.* 2012). The report found that many Germans believe that reconciling career and family is problematic and that having offspring isn't conceived as a central value of life.

In **France**, maternity leave is covered by the employer for the time of pregnancy and childbirth. The leave is paid for 6 continuous weeks before and 10 continuous weeks after childbirth at 100 percent of earnings for first and second children, for 8 weeks before and 18 weeks after childbirth for a third child, and for 12 weeks before and 22 weeks after confinement if multiple births (e.g., triplets). In addition, in cases of problematic pregnancy medical status, the leave covers 2 additional weeks. The extended maternity leave linked with working status has accompanied increasing female participation in the labor force. As of 2005, the state acknowledged the need to support employers in replacing employees on a temporary basis during the maternity leave, in order to make sure that the pregnant woman would return to work after her maternity leave period ends (Labour Code Art. L. 122 and law of 25 July 1994 and Article 60 of Law No. 2003–1199 December 2003 relating to the financing of Social Security). Paternity leave, as of January 2002, was extended from 3 to 14 days of paid, job-protected leave and to 21 days in the case of multiple births. The cash benefit equals the maximum benefit covered under social security. Parents are free to choose to take the leave at the same time or sequentially and have the right to return to the same or similar job at the same pay.

In **Denmark**, the maternity leave period was extended from 32 to 52 weeks in January 2002 (Bertelsmann Foundation 2003: 36); however, such reform cut the original one-year parental leave. Of these weeks, 4 must be taken during the pre-maternity period, while the father may take 2 weeks during the childbirth period and the first 14 weeks after birth are taken by the mother. The remaining 32 weeks may be "shared freely" between the father and the mother. Parents are allowed to take up to 32 weeks of parental leave after the maternity period, of which 8–13 weeks may be used until the child reaches the age of nine years (Bertelsmann Foundation 2003: 36). Parental leave lasts until the child is up to age 8 for a period that must not exceed 52 weeks. Parental leave may include uninterrupted periods of leave of at least 13 and at most 52 weeks, but these should be voluntarily approved by the employer. The fact that an employee makes use of the entitlement to childcare leave does not constitute a valid reason for dismissal, and the period of leave counts towards the employee's continuous length of service. The rationale behind this entitlement is to encourage temporary replacement in the jobs left vacant during the periods of leave.

The **Netherlands** offers a 16-week maternity leave at 100 percent wage replacement to a specified maximum. Unemployed women have the right to a lower benefit. Pre-maternity leave is taken 4–6 weeks prior to birth and maternity leave for 10–12 weeks following birth. A reform in 1997 extended the duration of parental leave with an additional 4–6 months' full-time unpaid, job-protected leave or its equivalent. Parental leave may be taken until the child is eight years old. In the case of multiple births, there is a separate entitlement for each child. It should be noted that in the Netherlands, parental leave is based on individual entitlement and therefore is not transferable between parents. Fathers are offered two days' paid paternity leave. Parents may also use up to ten days' family leave per year and an additional two days of emergency leave. As of January 2001, employees are allowed to set aside up to 10 percent of their gross annual income and/or the equivalent number of working hours to finance a leave of a maximum of 12 months.

Family and child allowances

From the 1950s through the 1980s, **Denmark** restored its universal approach, convinced of its rightness for child allowances. Child allowances are presumed to be larger for children up to age 2 than for children aged 3–6. These allowances mark Denmark as one of the most generous European countries in this area. These allowances are job-wage earning related, covered, and untaxed by the government; special allowances are awarded to student parents. In addition, non-custodial parents can be socially assisted by a separate grant (child support) covered by the local social welfare. Payments are in advance, semi-annual, and continue to age 18. In cases of children deprived of support because the supporting parent has had an occupational accident or illness, families may take survivor benefits for children under age 18.

In **Finland** parents may receive child allowances until a child reaches the age of 17. These allowances are graded according to the number of children in the family. In addition to the child allowance, a maternity care package is given to every mother who wants it. The package contains the requisite baby needs during its first year of life and encourages all pregnant women to visit maternity clinics for guidance and advice; visiting times are arranged in the best interests of all mothers and babies. The purpose of maternity clinics is to safeguard the welfare of mothers and unborn children. In addition to the child allowance, there is a separate grant for lone parents who are divorced or separated, or who have a child out of wedlock. Other Child Conditioned Income Transfers are provided by the social security system. This allowance targets child survivors of employees or spouses caring for children under the age of 18 (or 20, if a student).

These are universal benefits that help low-income orphans (Finnish Ministry of Social Affairs and Health 1999).

In **Germany**, cash benefits and tax benefits for children were included in the family benefits package (*Kindergeld*) in the mid-1990s. Families with one, two, or three children are granted a universal child benefit of €154 per month; for each child thereafter the amount increases to €179. Youth up to age 18 and in cases of unemployed youth age 21 or students until age 25 are allowed to receive benefits with a phase-in for birth years 1980–1982. In cases of disabled youth unable to work, there is no limitation to such allowances. The benefit may also be claimed for foster children and for grandchildren. In addition to the child allowance, there is a supplementary income-tested child benefit (*Kinderzuschlag*), which is paid to parents who are unable to cover their children's living expenses, if the benefit would prevent the family from having to apply for social assistance for the children. Child tax allowances or credits are wage-related tax deductions for childcare and childrearing costs (*Elterngeld*) from birth to age 14, receivable by employed lone parents and dual wage-earning partners. All other parents may receive the tax relief for children aged 3–6 (Federal Ministry for Family Affairs, Senior Citizens, Women and Youth 2007).

In addition, families who receive one of the child tax allowances or credits and have a child attending boarding school or pursuing vocational training, are entitled to a tax exemption to help cover the cost of education (Federal Ministry for Family Affairs 2007). Family allowance-related reforms in Germany also include pension credits to employed mothers or fathers who remain at home for up to three years with a child; the pension credit equals up to three years of work at average wages.

In cases of children deprived of support because the supporting parent has had an occupational accident or illness, families may take survivor benefits for children under age 18 under the social security system, and under work injury (workers' compensation) insurance. Accordingly, there are dependent child benefits under unemployment insurance and unemployment assistance. Child support is offered in grants for lone parents who are divorced or separated, or have a child out of wedlock (*Unterhaltsvorschuss*).

In **France**: The basic family allowance is a cash benefit provided regardless of income for each child beginning with the second up to age 19 (and to 22 under certain circumstances) and varying in amount by the child's age. These allowances are job-wage earning related, covered, and untaxed by the government. In addition, there are other categorical allowances, including social assistance cash benefit, available on a means-tested basis to those aged 25 and older, or those who are younger and have a child in care (RMI) (Richardson 1994), and childcare tax credit to offset some of the costs of childcare for working parents. Child support is offered in grants for lone parents and survivor benefits for children under age 18.

From the 1970s through the 1990s, the **Netherlands** introduced major child and family benefits that resulted in universal, government-financed allowances for children. In a reform made in 1994, for children born in 1995 or after, allowances are calculated annually to cover the real costs of rearing a child in each of the three age groups (up to 6 years, 6–12 years, and 12–18 years). These allowances are job-wage earning related, covered, and untaxed by the government, and can be extended to age 24 if the youth is in vocational training or further education and not entitled to student grants (Immervoll *et al.* 2000).

Several child-related tax benefits are offered to families, including additional credit for low-income families who have children 16 years or younger living with them. The lone-parent tax credit is for single-parent families whose children are under age 27 and living with the parent; there is a tax credit for couples with children aged 12 or younger living with them. Several child-related tax reductions exist. The survivor benefit under social security is paid for full orphans; in cases where children are reared by their widowed mother, the benefit is based on the mother's income-test, not the child's.

Public support accounted here only concerns public support that is exclusively for families (e.g., child payments and allowances, parental leave benefits, and childcare support). Spending recorded in other social policy domains such as health and housing support also assists families, but not exclusively, and is not included here. Data on tax breaks towards families are not available for Chile, Estonia, Greece, Hungary, Israel, Slovenia, and Turkey, and non-OECD countries (OECD 2007b).

The multi-option childcare benefits and the maternity-parental leaves and benefits are considered responsive to the high labor force participation rates of mothers. Family planning reforms in French, Finnish, and Danish family policy in extending periods of maternal and parental leave that may be taken after the

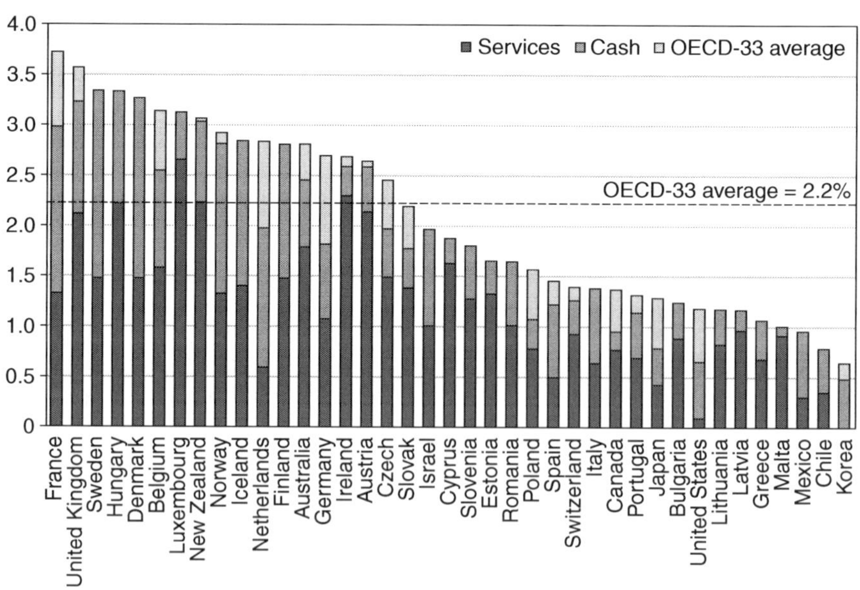

Figure 5.3 Public spending on family benefits in cash, services, and tax measures, in percent of GDP, 2007 (source: Social Expenditure Database [2010]. Online. HTTP: www.oecd.org/els/social/expenditure and ESSPROS 2010. Online. HTTP: www.oecd.org/els/familiesandchildren/37864391.pdf).

expiration of maternity leave, aim to improve the reconciliation between work and childbearing and care. However, in Germany and the Netherlands child allowances and maternity leaves do not encourage low-income parents to care for their children themselves—and hence to leave the labor market, with the risk of being trapped in unemployment and poverty, while the Nordic countries have strongly attempted to reduce social and gender inequalities.

Comparative analysis

Table 5.4 presents aggregated descriptive statistics presented in this chapter of ART and immigrant integration in the states under investigation. As seen, as with assisted reproduction technologies, the policies intended to regulate their use were developed only recently. Consequently, data on ART exist only for the past few years, constraining the ability to produce a more comprehensive statistically meaningful analysis. The data presented in Table 5.4 refer to the year 2007, and although the data do not allow for within- and between-country analysis over time, they allow us to compare among different states and to test for a relationship between ART policies and immigrant integration. In this respect, we see added value in presenting what we believe is indicative of a pattern in policy-making across countries in Europe.

Welfare-state type: Relying on Esping-Anderson's categorization of three levels of the welfare state, in ascending order we rank the states as follows: 1 = liberal welfare states (lowest level); 2 = conservative welfare states (intermediate level); 3 = social democratic welfare states (highest level). As discussed earlier, whereas the social democratic Nordic states of Sweden, Finland, and Denmark represent the highest form of welfare state, the conservative welfare states of the Netherlands, France, and Germany are ranked as intermediate welfare states.

MIPEX: depicts the Migration and Integration Policy Index developed by the British Council and the Migration Policy Group. The index allows for a comparative assessment of the degree of legal equality of immigrants across 25 EU member states. The index covers 140 indicators in six areas: access to nationality, long-term residence, anti-discrimination, family, labor market access, and political participation. While the social democratic welfare states rate relatively high on the index, moderate and low level welfare states are positioned lower.

National spending on ART: This variable depicts the state's annual national spending on assisted reproduction technologies as a percentage of the GDP. Table 5.4 shows that, as might be expected, the highest ranking welfare states (the generous Nordic welfare states) are the states that invest the greatest portion of their GDP in ART, with Denmark and Finland investing 0.06 percent of the GDP and Sweden investing 0.05 percent compared to the states ranking 0.02 on the welfare-state scale (France and Germany 0.04 percent, Netherlands 0.03 percent) of its GDP in ART. These figures give a crude estimation of ART public spending which may be considered high in terms of cost-effectiveness analysis since the costs of pregnancy care, neonatal care, mortality or morbidity associated with ART is not included (Sunde 2007; Chambers *et al.* 2007).

Table 5.4 National spending on ART, family benefits, and immigrant integration in seven West European states

Country	*Welfare-state type*	*MIPEX index 2007*	*State coverage of IVF*	*National spending on ART 2007 (as percent of GDP)*	*EMP*	*Public spending on family benefits*
Sweden	3	88	1	0.05	0.72	3.35
Finland	3	70	1	0.06	0.74	2.83
Denmark	3	51	1	0.06	0.78	3.28
Israel	3	–*	1	0.05	–**	1.99
Netherlands	2	68	0.50	0.03	0.77	2.84
France	2	55	0.50	0.04	0.83	3.71
Germany	2	53	0.50	0.04	0.87	2.71

Notes

* No available data for Israel as the variable was calculated for EU countries alone.

** No available data for Israel as the variable was calculated for EU countries alone.

Government coverage of IVF: This variable depicts the degree of coverage provided for women for IVF by the state, with three levels: full coverage = 1, partial coverage = 0.5, no coverage = 0. We see that states positioned high on the welfare-state scale also tend to provide greater governmental coverage for IVF.

EMP: This variable depicts the level of employment for non-EU immigrants. Relying on the Eurostat pilot study on the Indicators of Immigrant Integration, published in 2011, we use this variable as an indicator for measuring immigrant integration.

Public spending on family benefits: This variable depicts the state's public spending on family benefits in cash, services, and tax measures, in percent of GDP based on the OECD report of 2007 (2007b).

Summing up

In this chapter, we used the welfare-productivist model of Sweden and Israel to extend the comparative analysis to other welfare regimes and their policy convergence in funding assisted reproduction technologies to increase productivism compared to immigration policies, and that the general philosophy behind welfare regimes appears to be enduring.

Despite the present divergence of welfarist policies in several European countries, they all share a similar characteristic that has not yet been discussed within the extensive literature on welfare typologies—eugenic productivism. While comparative research on welfare regimes is often restricted to structure, and solidarity trends in state programs followed Esping-Andersen's typology (1990, 1999) differentiating welfare regimes into three principle types, namely, liberal welfare regimes, social democratic regimes, and conservative welfare states, in this chapter we suggest that all these welfarist models converge into the social democratic one by using the category of eugenic productivism. Recent developments in family policies and ART reveal a remarkable "welfarist" characteristic—driven by demographic concern for declining birth rate and increasing level of immigration in Europe—raising maternity and parental leave in general, extension of public childcare, and coverage for fertility treatments have finally been brought into being.

From the analysis of population policies, we saw that there is a general trend of convergence. That is, the higher the state is positioned on the welfare-state scale and, subsequently, the greater it supports assisted reproduction, the lower the immigration integration level of that state. This implies that in states that provide greater government funding for reproduction technologies such as IVF, not only does this funding naturally support the more productive segment of society (the career-oriented women) who require it the most—but this funding also likely comes at the expense of funding that could be allocated towards programs intended to enhance immigrant integration. The result is lower levels of immigrant integration in the countries supposedly most oriented towards social integration.

In terms of direction, we observed a pattern of increase in regulatory restrictiveness of integration and regulatory permissiveness in ART. As expected, the rise in restrictiveness is more reflected in generous welfarist countries such as

Finland and Denmark than for less generous, more liberal welfarist countries such as France and Germany and the Netherlands. This finding is in line with recent studies (Koopman 2010; Malmberg-Heimonen and Julkunen 2006) indicating that there is a withdrawal of welfare regimes and governments from the basic tenets of welfare provisions due to the fact that such regimes facilitate welfare-dependent (less productive) immigrants' access to welfare-state arrangements. However, this chapter suggests that the commitment to social spending and social transfers to working "productivist" couples and families has not ended. Moreover, it is argued that liberal democracies' commitment to uphold multicultural and free-market oriented values has shown some return to welfare nationalism and productivism; a productivism that is not universal. Although these countries need an increase in population and productivity, they have become increasingly reluctant to allow more immigrants to cross their borders. In an attempt to meet these needs, said countries encourage "the right" productive national stock to reproduce by extensions of access and coverage for expensive fertility treatments.

In terms of the increasing homogeneity of population policies in our research, the most significant conjucture is on family planning policies. There is a general trend of catching up by conservative welfare countries for such policies. The explanation lies, in Foucauldian terms, with the national construction of the biopolitical problem shared by European countries—the fear of low birth rate and infertility among productive national stock. Current trends in birth rate advocate a pronatalist policy, which faces the challenges of declining marital stability and delayed entry into parenthood. This development of delayed family-building and a reduced willingness to have children can be seen as a productivist effect of the difficulties in the labor market, especially during economic crises, as unemployment affected mainly those groups that now demonstrate very low fertility rates. In comparison to Nordic countries such as Finland, Sweden, and Denmark, the German conservative welfare-state regime that upholds traditional family values and which (until recently) has done little to support the compatibility between work and female employment, could explain why West Germany has been the forerunner in the patterns of very low fertility, high childlessness, and low transition rates to higher-order births.

Economically, after World War II, the immigration policy of these countries shared the same rationale in their immigration policy: importing immigrant laborers to boost production. Even in countries such as Finland and Denmark that aimed at extending immigrants' equal membership and the right to full participation in the community through labor market measures such as job training, education, increasing language skills, or support for self-employment, it is questionable whether they actually led to the real exercise and practice of full social citizenship. Indeed, integration policies directed towards ethnic groups have succeeded only in the creation of low-paying jobs within the public sector, resulting in a class of working poor.

The multi-option childcare benefits and the maternity-parental leaves and benefits are considered to be responsive to the high labor force participation rates of mothers.

Following the specific characteristics of population policy—such as birth rate decline, immigration policy, ART, and family allowances and leaves, despite variations among Sweden, Israel, Finland, Germany, France, and the Netherlands, similarities can be explained by the productivist eugenic logic of each national regulation. It is claimed that for more liberal welfarist regimes such as Germany and France it took more time to follow the same track as did the forerunner social democratic countries such as Israel, Sweden, and later Denmark, Finland, and the Netherlands. The political foundations for the Nordic welfare states have been closely linked with the Left—especially Social Democracy and Left political institutions.

Finally, attitudes tend to become more permissive regarding the integration of new technologies that gradually takes place in European states. Favoring permissive policy designs in the field of ART marks the trust put into biomedical and scientific progress as a beneficial measure for raising a nation's productivity. Social Democratic parties play a pivotal role in addressing a more permissive regulation of ART, including minimizing state intervention, increasing coverage for fertility treatments, and extending eligibility criteria relating to age and gender for native citizens.

In the absence of comprehensive ART legislation, Finland belongs to the permissive category by default with low integration. It seems that Finland created a friendly ART environment especially for practitioners of ART who are autonomous, use the widest range of techniques, including egg and embryo donation, surrogate motherhood, and stem cell research, in comparison to other Nordic countries. Much of the provision of ART in the country is regulated by the medical community, allowing access for single women and lesbian couples. Denmark allows egg donation, stem cell research, and preimplantation diagnosis. The autonomy of medical practitioners of ART is thus quite high. The Danish Act on Artificial Fertilization limits access to infertility services to heterosexual couples, but single women and lesbian couples have access via what could be characterized as a loophole in the Act. In practical terms, then, the provision of ART in Denmark is fairly liberal. However, since the level of access to infertility services is only medium *de lege*, the country belongs to the intermediate category of policy design. The policy designs in the Netherlands and France fall into the intermediate category in which access to ART treatments is limited to stable heterosexual couples. Germany has entered its third generation of ART policies through several redesigns. Similar to other countries, the German ART policy design has undergone a gradual transformation from a fairly restrictive design to a relatively less restrictive one.

Drawing on the comparative policy analysis, an important direction for future research is the comparison of our research results with the development of population policies' outcome data. Moreover, this research has also pointed out that what is significant is not the evidence of convergence but rather the explanation of convergence. The future challenge lies in investigating the causes of convergence that rest on various assumptions about the behavior of policy actors within each state.

Conclusions

> ... finally, you will find it in the racist biologists and eugenicists of the late nineteenth century. It is a sophisticated discourse, a scientific discourse, an erudite discourse spoken by people with dust in their eyes and dust on their fingers, but it is also—as you will see—a discourse that certainly had an immense number of popular and anonymous speakers. What is this discourse saying? Well, I think it is saying this: no matter what philosophical-juridical theory may say, political power does not begin when the war ends.
>
> (Foucault 1976[2003]: 50)

As we have remarked in this book, social scientists tend to view eugenics as a purely historical phenomenon whose relevance to current political and social circumstances is minimized. As can be expected, eugenic activists publicly arguing for eugenic policies are unlikely to appear. Indeed, the eugenic discourse might be defined as politically incorrect. However, the question is whether beneath political correctness a renewed version of eugenics is emerging.

This book, as we have seen, argues that under current socio-economic and cultural contexts, overt eugenic policies become attractive for policymakers.

We have claimed that in current times eugenic strategies are used under the guise of productivist strategies by nations interested in preserving a two-track goal: first and foremost preserving their ethno-national identity, or pretended national identity and, following suit, sticking to productive competitiveness in an interconnected world.

While the countries we analyzed in this book basically adhered to political liberalism, to liberal economics, and profess to reward productivity, there seems to be more than a coincidental relationship between the "productive" factor, which is universal and apparently non-national, and the ethnic national stock factor. In other words, the ideas of productivity and protection of the national stock are bound together, augmenting each other.

What we have traced is how in the countries analyzed, which could be defined as national, welfare, and productivist societies, the "right people" to support are the productive people, ethnically defined, and the welfare system serves their purposes. Following this point we have suggested and stressed that the expense of new reproductive technologies, the increasing role of women in the workplace, and the consequent ability to subsidize childbirth through parental leave policies, provides an incubator for eugenics.

This book has entered into one of the major debates in contemporary bioethics about the social and ethical impact of the new genetic technologies in which eugenics has increasingly come to occupy a central position, and has suggested a provocative interpretation based on empirical findings. We have suggested that the controversies surrounding the continuation or resurrection of eugenic population policies from the first half of the twentieth century to current times obscured the unavoidable relation between bioethics and politics. What began as racial eugenics and shifted into productivist eugenics akin to national welfare societies, reappears nowadays guised under genetic progress, the base of biopolitics, which slams bioethical discourse.

It is claimed that contemporary bioethics should go beyond the moral dilemmas about life and death that characterize traditional moral philosophy. Contemporary bioethics instead takes the path of critical examination of practice that might otherwise be taken for granted.

Regarding this last point, we argue in this book that the constitutive effects of bioethical discourse can be seen from the structuring of power relations. This consideration makes it obviously necessary to make use of Foucault's ideas in order to underpin bioethics as a source of critical thinking. Using Foucault's ideas on knowledge and power, we have developed a framework for the understanding of population policy in welfare states. Indeed, eugenics as a source of knowledge can be elicited to "help" decision makers to define their goals as rational judgments and to remove barriers to that logic. Identification and exploration of power relations enable bioethicists to deal with the way power abuse, dominance, and inequality are enacted and reproduced in the social and political contexts.

Thus, much of this research draws on biopolitics as its theoretical foundation, suggested by Foucault as the aspect of power relations that affects issues concerning the life of a population. Such aspect should be part of bioethical thinking in eugenic-based population policy.

Using Foucault's conceptualization of biopolitics reveals the not-so-visible techniques of biopolitics that surround population policies in welfarist countries through the mechanisms of biopower. Reflecting on the power/knowledge nexus around the discourse of the "life of the population" includes reflection on possibilities as well as constraints and thus gives rise to the exclusionary technique that was identified by Foucault as biopower.

Indeed, drawing on Foucault's ideas, analyzing bioethics and power in modern society reveals different forms in which people are given new possibilities for choice under the guise of "free choice." In practice, in the rational planning of the population in welfare nations, eugenic techniques can be understood as the offering to individuals of possibilities to enhance their life choices including reproductive choices. At the same time, eugenics proves to offer useful knowledge for improving population life while excluding "degenerate and unproductive" groups within a given population. The two areas of inquiry—

possibilities (inclusion) and constraints (exclusion)—constitute a biopolitical form of bioethical critical thinking of eugenic-based population policy. Population planning has become the core exercise in the nation-building project where the national interest lies in developing the economic (productive) and demographic (reproductive) profile of nations. Thus, Foucault's extended account of biopolitics is used in the welfarist discourse by simultaneously transforming reproduction into an object of study and the "productive citizens" into a subject of power.

In the light of power relations, the opening up of bioethical debate over eugenic practice from a biopolitical approach addressed the biopolitical role of national identity, assuming that the life of a given population should override the individual's rights. In the practice of population management, individuals are given possibilities for reproductive choice under the guise of individual right, where these possibilities are constrained by practitioners' authoritative power to prescribe what is best for the population as a whole.

Moreover, Foucault's genealogical method helps us to understand the wider context of relations in which eugenic forms have been deployed in population policies. Genealogical method offers to explore the context of knowledge in terms of social practices that sustain it via power relations. The genealogy of eugenics as a form of population policy in welfarist nations aims at describing the conditions of emergence of the present form of eugenics associated with the new reproductive technologies and their connections with wider social practices such as population policy. A genealogical analysis of power relations associated with eugenics yields special insight for people's-home welfarist democracies by looking at how welfare states in which nationalism facilitates exclusion of marginalized groups under the guise of welfarist state commitment to equal access to state provisions.

Thus, this book has offered an account of how eugenics was being incorporated into population-based analysis and became politically prominent in various political settings. In other words, the history of eugenics as a scientific discipline and as an ideology is instructive for those concerned with the bioethical issues surrounding the new genetics. In addressing these issues, this book analyzes power relations that enable eugenic forms to be accepted by welfarist people's-home nations. Following the "continuity hypothesis" from welfarist productivism to ART presented in this book, the recent increase in availability of genetic technologies may provide a surreptitious way for governments to control the demographic makeup of their countries. Such continuity is explained by the disciplinary history of eugenics.

Based on the twofold nature of eugenics as a scientific research field and public policy, the two discourses conflated in examining the nature of eugenics: the scholarly discourse situating eugenics in the history of science and aiming at understanding the intellectual development in scientific and ideological contexts, and the political discourse that we set out in this book to compare the historical and the current conjunctures between eugenics and new genetics in population policies in welfare states.

In general, it is argued that eugenics failed to become an influential or legitimate scientific discipline. In both anthropology and human genetics, eugenics did not fit into the paradigmatic scientific framework. Within the anthropological field, eugenics aimed at establishing a causal link between race and hereditary biology, and in turn provided a distorted judgment about how inferior races ought to be treated within society. However, the reorientation in anthropological studies of human races signaled the crystallization of a paradigmatic consensus that the concept of race had no biological validity and that the many intermediate grades of variation between races made the distinction between them an arbitrary one. In the case of human genetics, eugenics refused to regard the model of variability in large populations as compatible with Mendelian heredity, and instead showed how artificial selection should be used to remove the unfit from the social body in the name of improvement of the human race. However, when the modern synthesis established itself as a permanent feature of the genetics landscape, eugenics could no longer provide a conceptual territory within the biological sciences in which the discipline could operate. Finally, during the 1930s there was considerable agreement among geneticists that the modern synthesis legitimately constituted the paradigmatic core of a genetic discipline.

Relating these considerations to the ideological discourse, these findings seem to be consistent with a strain of thought that views eugenics as a discredited and illegitimate discipline. Within the ideological discourse that we have outlined, eugenic concepts could not support claims for a coherent ideology. Degeneration as a core component of ideology took a peculiar form in eugenics, in that it insisted on an essential identity of biological race and degeneration. However, the 1930s' leading political ideologists had shifted from claims about the presumed link between degeneration and biology. Social-environmental concepts such as youth, class, nation, and imperialism were integrated by totalitarian regimes such as fascism and communism into a goal-directed and coherent ideological "core." The general trend of transformation from racial-biological concepts to social-environmental ones maintained the unity and relative stability of degeneration as an ideological core. This trend paved the way to the productivist conceptualism of degeneration.

Although eugenics did not become as coherent an ideology as fascism, Nazism, and Leninism, it has not lost its ideological doctrine of degeneration. Eugenics' looser ideological structure gave eugenics its flexibility and ideological adaptation to serve more coherent ideologies such as nationalism by offering an appealing set of social-political exclusionary practices. However, such a conclusion of the apparent failure of eugenics as both scientific and ideological disregards eugenics' "power" to respond to age-old problems in the eyes of policymakers. As we have noted and stressed, this type of "exclusionist" inclusionism conditions entrance into the collective identity. It is not closed to a specific ethnic group, though it is assumed that the specific ethnic group's cultural and productive values should prevail. This is done by promoting fertility of the hegemonic group and national integration of productive foreigners.

Indeed, eugenics has not lost its rational logic to manage population, especially after World War II, when "unproductive people" became profoundly damaging to nation-state survival.

It is true, though, that eugenics was discredited after the war for its presumed association with Nazism. However, as we have shown, eugenics fit social democratic or labor movements' productivist ideologies and were rather suited to welfare social democratic Lutheran countries rather than to Nazism. As we have suggested, while Nazi racism gave priority to race over productionism, social democratic movements gave priority to productivist inclusion over race.

In this sense we suggested that eugenics became a service of social democracy ideology rather than as a coherent ideology in itself, and consequently survived the transition from "racial hygiene" to "national hygiene" politics. The historical-institutionalist analysis presented in this book implies that the adoption of eugenic-based policies should have had a significant impact on the development of population welfare-state policies. The historical look at Swedish, Danish, and Zionist labor and social democratic movements as well as of the Finnish state during the 1930s and 1940s reveals that despite fundamental differences, elements of eugenic continuity persist by surviving the transition from "racial hygiene" to "national hygiene" politics. This transition opens the path for the shift from eugenics to genetic research and its role in developed Western democracies in current times.

Thus the historical analysis presented in this book locates this shift from "racial" eugenics to "national-welfare" eugenics during the 1930s, where eugenic policies could no longer be understood independent of socio-economic factors. As we have suggested, eugenic strategies were favored by social democratic and labor movements to preserve their national identity, which does not necessarily imply pure racist exclusionism.

Indeed, national communitarian states or nation states did not necessarily endeavor to promote the "racial purity" of the nation. Although racist states like Nazi Germany were characterized by the search for racial purity, other national communitarian states displayed "national productivist" purposes. In some states, however, the starting point could be ethno-exclusionist, although they turned productivist. That does not mean that the ethno-exclusionist factor disappeared. However, it means that the productivist argument becomes dominant. In Sweden, Denmark, Finland, and Israel, until the 1920s "racial hygiene" was regarded as an analytical category to legitimize and reinforce the social process through which individuals or groups are constructed as being members of an ethno-homogeneous state. However, this "collectivist" understanding of eugenics paved the way to another "collectivist" type of eugenics—a social-productivist one. In Finland the racial hygiene promoted by the Swedish minority became a social-productivist one at the service of the whole nation.

Indeed, in the noted countries, health practitioners and scientists posited a common ideology of productivism and welfare, integrated with a national concern that the population whose members were mentally or physically sick were profoundly damaging to the survival of the nation state. The Zionist solutions luckily did not come close to those of the Swedish, Danish, and even Finnish sterilization policies, but managing the Jewish mentally ill in Palestine broadly fit into the eugenic framework of preventing the "problem of unproductive people" through immigration policy, or eliminating it by means of exclusionary immigration and repatriation. These solutions were complicated by a lack of medical facilities, limited funds, the presence of an imperial power, and the internal political conflict continually brewing in the land. In all these countries, health practitioners and scientists posited a common ideology of productivism and welfare, integrated with a national concern that the population whose members were mentally or physically sick were profoundly damaging to the survival of the nation state. Yet observers have good reason to suspect that the synthesis of welfare productivism and eugenics aids those who want to protect the nation by excluding those that are ethnically different. Since modern democracies find it difficult, if not impossible, to rationalize the goals of eugenics that discriminate on the basis of race or ethnicity, the national productivist formula may come to their aid. Finland, Denmark, and Sweden were able to enact such policies by reconciling their national agenda, which aimed to raise the fertility of their population with a democratic social agenda that had as its goal equal distribution of benefits to increase the welfare of the nation's citizens and residents more generally. Again, as noted, the Finnish, Danish, and Swedish identities are not fully "closed" identities. However, in the Finnish, Swedish, and Danish cases, inclusion into the ethnos implied cultural assimilation to productivist terms. The case of Israel is more complicated because inclusion depends on Jewishness. Nonetheless, the productivist factor is still prominent. The state indeed, paradoxically despite the increasing power of religious conservative parties, supports positive eugenics among productivist Jews, including lesbians, homosexuals, and single mothers. Indeed the Jewish and productive factors are tied together. In the past the productive Jewishness was associated with Ashkenazi Jews. In current times it is associated with all Jews; however, positive eugenics as noted is used in order to raise fertility among all its productive Jewish population.

In other words, the fear of societal degeneration has not disappeared. Welfare states may encounter a shortage of resources for health and welfare and therefore may become more interested in eugenic policies that dictate the national agenda to marginal social groups in the industrial society. Indeed, developed welfare nations, challenged by economic competition and the pressure of immigration, might aim to reduce the costs of welfare dependency by using genetics to ensure that as many children as possible are born with genes that enable people to live full and independent lives with no need for welfare dependence. As we have maintained, the biopolitical discourse on the possible continuity of eugenics in contemporary human genetics has stressed the transition to economic framing of eugenic-based population rather than overtly racial policy, which secured the prominence of eugenics in politics.

Thus the biopolitical discourse offers to extend the extensive literature on welfare-regime typologies by stressing the category of eugenic-productivism.

While comparative research on welfare regimes presents a divergence of welfarist policies in several European countries, this book highlights the role of eugenic-productivism-national relations in circumscribing bioethical debate. It stressed the continuity of eugenics in contemporary reproductive genetics and it presents a bioethical critical thinking that was not previously discussed.

Notwithstanding the historical and institutional differences between diverse types of welfare systems offered in the literature, what the story of our case studies unfolds is the extent to which welfarist regimes converge into the eugenic-productivist one.

Currently, it is suggested that the ideational features of eugenics thus securing a place for path dependency have been delineated as factors influencing welfarist policy and politics. In this book we claim that population policies are designed in a context where citizens and government agencies share a commitment to welfare-state policy choices, thereby limiting the range of other possible trajectories. Within the context of national welfarist productivism, the eugenic ideas take a particular form that concerns population and fertility subsiding. Although national response to changing demography varies with the institutional pattern of welfare regimes, national governments use a eugenic-productivist approach to harmonize their social programs by placing productivist citizens' fertility and reproduction on their agenda and their regulations. Such an exclusionist approach to traditional universal welfare values is reflected in providing incentives and funding to encourage the reproduction of native-born or long-time settlers.

Our evaluation of the welfare-state model allowed us to assess the link between eugenics and socialist reformist practice. National communitarian states or nation states needed the "welfare-productivist-eugenic" link in order to provide a race-neutral differentiation between those people who needed to be preserved and the people who did not. We suggested that genetics is still being used for social engineering in welfare states and that the resultant exclusion of individuals and communities is a problem demanding social-political attention. In tracing exclusionary rationing to national productivism, both the Israeli and the Swedish welfare states clearly differentiate according to group membership. At one extreme are citizens who benefit from the generous social subsidies for new genetic reproductive technology. At the other extreme are immigrants and ethnically different citizens, who have less access to social transfers and employment. Welfare states provide generous benefits such as reproductive technology and child allowances to members of their national stock, at the same time hardly providing benefits to those who do not share in ethnic membership.

This book puts forward the claim that governments in European countries have followed a similar track as that of Sweden, Denmark, Finland, and Israel by using biopower rationalization of eugenic productivist policies to cope with demographic crisis. The demographic challenge includes changing family structure such as increasing numbers of consensual marital relationships, couples producing fewer children, and more children being raised by households of only one parent. Moreover, declining birth rates have led to an aging population and more mothers are working for pay, placing pressures on both employers and governments to find solutions to childcare and family leave requirements. These challenges question the basic tenets of the welfare states to cope with reducing income taxes, social spending, and "welfare dependency." The major dimensions offered for analysis of the current eugenic-productivist-based population practices include those aimed at raising birth rates, targeting productive subgroups in society including restricting foreign immigration, funding and accessibility to assisted reproduction technologies, and employing expensive family planning programs.

Current trends in birth rate advocate pronatalist policy targeting certain subgroups. While pronatalist policy aims at struggling with the current trend of delayed family-building and a reduced willingness to have children, namely increasing all births, eugenic policy aims at increasing "worthy" or "good" births. This can be seen as a productivist effect of the difficulties in the labor market especially during economic crises as unemployment affected mainly those groups that now demonstrate very low fertility rates. In comparison to Nordic countries such as Finland, Sweden, and Denmark, the German conservative welfare-state regime, which upholds traditional family values and which (until recently) has done little to support the compatibility between work and female employment, has become associated with low fertility, high childlessness, and low transition rates to higher-order births.

Indeed, integration policies directed towards ethnic groups have succeeded only in the creation of low-paying jobs within the public sector, resulting in a class of working poor. As we have seen from the brief overview of immigration policies in the past, during times of economic recessions, unemployment impacts immigrants more than nationals. Immigrants tend to be associated with greater risk of unemployment and work stress because the level of the chief forms of social insurance against sickness, disability, unemployment, and old age are tied to the level of earnings, while those who cannot participate receive less generous flat-rate compensation. In practice, chronic unemployment relegates individuals to the marginal area of the social security system. Economically, after World War II, the immigration policy of these countries shared the same rationale of immigration policy: importing immigrant laborers to boost production. Even countries such as Sweden, Finland, and Denmark aimed at extending immigrants' equal membership and the right to full participation in the community through labor market incentives such as job training, education, increasing language skills, or support for self-employment. The question of whether any of these actually lead to the real integration and exercise of full social citizenship is still open. In any case it seems quite clear that investing in fertility of the social strong "middle classes" becomes a strategic option that cannot be dismissed.

The multi-option childcare benefits and the maternity-parental leaves and benefits are considered to be responsive to the high labor force participation rates of mothers. Recent reforms in German, Finnish, and Danish family policy in extending periods of maternal and parental leave that may be taken after the expiration of maternity leave aim to improve the reconciliation between work and childbearing and care. However, in France, Germany, and the Netherlands child allowances and maternity leaves do not encourage low-income parents to care for their children themselves—and hence leave the labor market—with the risk of being trapped in unemployment and poverty, while the Nordic countries have intensely attempted to reduce social and gender inequalities.

Within the field of ART, European states gradually favored permissive policy designs in terms of the nature and scope of regulation of reproductive technologies, public coverage, and rules of access, stressing that biomedical and scientific progress is beneficial for raising labor productivity. Such an approach bolsters the idea that state intervention should be kept at a minimum and coverage for fertility treatments should be increased. Finland belongs to the permissive category by default. The practitioners of ART in Finland are able to use the widest range of techniques among the Nordic countries, including egg and embryo donation, surrogate motherhood, and stem cell research. Furthermore, infertility services are accessible for single women and lesbian couples, making the level of access high as well. The Swedish ART policy design has undergone a gradual transformation from a fairly restrictive design to a permissive one. Both autonomy of the medical practitioners and access to ART point at comprehensive and liberal access to reproductive technologies incorporated in the current Swedish legislation. The only remaining restriction is the ban on embryo donation. Although surrogate motherhood is not explicitly prohibited in the Swedish law, surrogacy is not practiced in the country. Since July 1, 2005, lesbian couples, whether cohabiting or living in a registered partnership, are allowed by law to have access to infertility services. Denmark allows egg donation, stem cell research, and preimplantation diagnosis. The Danish Act on Artificial Fertilization limits access to infertility services to heterosexual couples, but single women and lesbian couples have access via what could be characterized as a loophole in the Act. The fact that Denmark reached the highest level of ART treatment per million people per year in Europe—almost 5 percent of all children conceived in Denmark—forced the government in May 2010 to cut all reimbursement for ART. The position of Germany, France, and the Netherlands is more intermediate towards ART regulation. In Germany, the eligibility for ART treatment is limited to married couples and heterosexual couples cohabiting for a minimum of two years. The use of donated gametes is restricted to married couples only. Sperm donation is permitted, although payment for such donations is prohibited. Women, however, are constrained from donating their ova. Furthermore, research related to embryos is narrowly defined and heavily regulated as well. However, it should be noted that the public health care insurance covers 50 percent of the costs for up to three cycles; in addition, coverage depends on marital status and age. Given the integration of ART policy into the health care system, France has a broader access to ART services than would have been expected from the broader welfare-state context in which it operates. However, access to ART treatments is limited to stable heterosexual couples. Same-sex female couples do not qualify for support for ART treatment under this paradigm. Coverage is de facto universal and all residents in France have the right to access ART-related services, with the restriction of medically justified age limits and following best practice established by the profession. Thus, universal access to ART is in contrast to the set-up of the wider welfare and social security system, which is characterized by a strong link between work position (and/or family state) and social entitlements. Comparatively, the Dutch regulation of ART also consists of an intermediate approach. The Social Democrats did use the window of opportunity to change the ART policy of the Christian Democrats to take a more permissive approach. Indeed, the return to power of the Christian Democrats in the fall of 2002 did not lead to withdrawal from the framework put forward by the secular coalition and imposition of their restrictive policy goals.

Following the specific characteristics of population policy—such as birth rate debate, immigration policy, ART, and family allowances and leaves—despite variations among Sweden, Israel, Finland, Germany, France, and the Netherlands, similarities can be explained by the productivist eugenic logic of each national regulation. It is claimed that for more liberal welfarist regimes, such as Germany and France, it took more time to follow the same track as did the forerunner "social democratic" countries such as Israel, Sweden, and later Denmark, Finland, and the Netherlands, and settled for a regime in which ART techniques are permitted, but heavily regulated and supervised by government agencies. The political foundations for the Nordic welfare states have been closely linked with the Left—especially social democracy and left-wing political institutions, or by Lutheran social democratic countries such as Finland that develop a welfare social democracy without a social democratic movement having a leading role.

In sum, as we prove in this book, the ideological subtext behind genetic research in Western democracies is related to a twofold strategy—productivity and protecting the national stock. Rather than opening the gates of immigration and implanting costly civic integration programs or universal welfare, the new trend in some Western European democracies is a marked preference for ART and family environments. The productivist underpinnings of these policies lead to the conclusion that the eugenic policies of the past are refurbished to provide responses to similar logic. With regard to the development of population policy, the four countries profiled here have followed different approaches but converge towards the same population policy mix: different in the details of emphasis and structure, but similar in their national productivist practice. High levels of economic pressures and declining TFR create similar problems and opportunities, which are dealt with in broadly the same manner in different countries, regardless of the differences in their welfare regimes. The debate about the connecting line between welfare and productionism and with the necessity of population growth that leads directly to eugenic politics guised as genetics in present days is what this book is about.

What is to be done? Civic integration? Eugenics?

There is no question that if Europe wants to keep growing economically it needs among other things more population. The countries we have studied in this book provide one way to cope with the twofold problem of productivity and protection of the national stock. Those are liberal countries, which, although paying lip service to multicultural ideology and liberal ideology, are increasingly preoccupied with the protection of what can roughly be defined as their national culture, or societal culture as defined by Will Kymlicka. One thing must be clarified. The very idea of protecting a national culture, let alone a national stock, which for many observers may represent the same thing, is problematic in modern democracies. Indeed, the very idea of protecting a nation-state culture in times of growing immigration may sound non-democratic. A demos (the people), according to Robert Dahl, should not be left to those who contingently make it up in a given time and place. In that case a demos could become a limited group that exercises brutal despotism over a vast subject population (Dahl 1989: 122).

Scholars such as Yasemin Soysal, for example, suggest that for years national belonging constituted the source of rights and duties of individuals, while the nationally circumscribed public sphere constituted them the locus of people's claims (Soysal 1996: 4). The new trends accompanying globalization helped to recast national citizen rights as global human rights. These rights thus are becoming more abstract and less related to national belonging. Other scholars such as Seyla Benhabib suggest that current nation-state democracy seems to be an anachronistic type of democracy. The reason is that while it might be true that the modern nation state remains a possible structural expression of democratic self-determination, the complexity of our social lives integrates into associations that lie above and below the level of the nation state (Benhabib 2002: 169). Following this line of thought, scholars like Tony Judt have criticized Israel, claiming that it has come late to the model of the nation-state "nationalism" (Judt 2003). Most of these scholars claim that a liberal open world is what is needed today, stressing that nationalism is passé in times of growing migrations.

However, world developments hardly confirm the forecast of those downplaying the very idea of nationalism. Despite claims in favor of multiculturalism and in favor of a post-national European citizenship, it seems that national identities are returning to public consciousness. As a result of the growing presence of Islam in Europe, it seems that more than ever there is a backlash in Western societies against multiculturalism. As German Chancellor Angela Merkel declared, multiculturalism in Germany has failed. A study of the Ebert Foundation indicates that nearly a third of respondents agreed that foreigners come to abuse the welfare state and that foreigners might overrun the country (Evans 2010). Several scholars would define this anti-multicultural flavor as a resurgence of racism. The growing presence and dangerous turnouts of "right-wing populist" parties in European or national elections even in liberal countries like Sweden, indeed, mark a tendency that led many observers to determine that Europe is clearly shifting to the Right. As Ian Buruma claims, "First in Denmark, then in the Netherlands, and now in Sweden, illiberal, populist parties stirring up fear of immigrants—specifically Muslim immigrants—have managed to gain enough power to set, or at least influence, their countries' political agendas" (Buruma 2010). In more senses than one, however, we are not witnessing a right-wing populist overtaking in liberal societies, but a shifting from liberal multiculturalism to liberal nationalism, and a new stance of Western population in favor of a synthesis between liberal rights and national culture. For the latter, the idea of national citizenship is becoming more important than ever.

Why are national identities back in fashion? Precisely because Europe as a political and cultural entity is rather weak and, as a result of that, feeling there is no definitive conception of what it is to be European. National identities, though disputed, constitute a decisive point of reference. Despite particular differences, most European countries have arrived at a middle ground where their national particularisms constitute local versions of a universalistic idiom of liberal democracy (Joppke 2010). This middle-ground narrative could be defined as a shared culture that cannot be equated to Habermas's constitutional patriotism, although it cannot be stigmatized as exclusionist. As noted, scholars such as David Miller and Dominique Schnapper emphasize that only a shared identity may set the underpinnings for the twin development of democracy and social justice. Schnapper links national solidarity with a common political project stemming from a political culture that public schools, for example, are in charge of promoting (Schnapper 1994: 41). This development means a lot with regard to one basic question: How European societies which are already multiethnic and multicultural will deal with immigration. Scholars such as Randall Hansen stress the importance of national citizenship and challenges from an empirical and normative way the trivialization of the idea of citizenship as advanced by post-nationalist scholars (Hansen 2009: 2). In other words, while liberal multicultural scholars suggest national identities should change in order to accommodate the pressing issues of society, nowadays we can perceive a comeback of those theories that stress the importance of the ultra-stability of national citizenship traditions.[1] Precisely because of the reasons mentioned, European countries are enticed to refurbish their mythical/invented national culture. As stated by Herman van Gunsteren, one consequence of the debate on integration has been a shift from the understanding of civility as a political concept to a social one, a trend encouraged by increased involvement of the state (van Gunsteren 2010). This "depolarization" of civility asserts that one has to be integrated first into society and local culture before one asserts oneself in the political arena. In more senses than one that means that a liberal system is not neutral among conceptions of the good.

Saying this, it should not be strange to hear that several liberal democracies that are welfare states and productive want to preserve the hegemony of their societal culture. For the latter, the preservation of that societal culture is not exclusionism but precisely the opposite. It ensures the benefits of society for all. As we have analyzed in this book, whether we like it or not, the question of "population analysis" is an integral part of this debate, which focuses in a central question: Whether these types of national societies can survive economically. Either they become more productionist through immigration or less productionist by closing the gates of immigration. The third option, as we have seen, could be divided into two-track strategies, which in some cases could fit each other.

The first strategy is civic national integration, which implies a policy of opening gates to immigration, although under very strict conditions of integration, which fall almost totally on the newcomer. The latter are required to integrate into the national culture and become productive individuals for the benefit of the community. The second track is that analyzed in this book, which is promoting the breeding of the productive nationals.

The first strategy of civic integration does not want to halt immigration, precisely the opposite. In the early 1990s it had been the joint stance of European states to sternly reject new labor migration. Closure to the outside was often taken as a precondition for accommodation of immigrants who had already been admitted. However, since the end of the 1990s most West European countries have opened their doors to the economic elite of globalization. Even Germany's new *Immigration Law* of 2004 offers permanent settlement visas from the start for highly skilled immigrants. Indeed, Europe was looking for talent and opening the market for newcomers with skills, although in small numbers in comparison with countries of immigration. Still we must emphasize that as Europe is permeated by right-wing populism, the imperative to keep the numbers small is still strong. As Christian Joppke has noted, two parameters have changed in European politics with regard to immigration since the mid-1990s. In the realm of ideas and practice a new consensus arose in favor of legal immigration, reversing three decades in which immigration was unwanted (Joppke 2007: 244). This trend did not change much even after acts of terrorism swept some countries in Europe. In 2005, as a result of terrorist attacks in Europe, political leaders became aware that civic integration policies for immigrants should become obligatory, and as defined by Desmond King this liberal contractualism between the newcomers and the state was clearly balanced towards duties of the immigrant rather than rights he gets from the state. According to King, thus civic integration is an instance, next to eugenics in the past and workfare today, of "illiberal social policy" in a liberal state. King's important insight is that such policies are not born of sources extrinsic to liberalism, such as nationalism or racism, but are inherent in liberalism itself (King 1999, cited in Joppke 2007: 249).

In this book, although we agree with this definition, we tend to link liberalism with civic integration and eugenics with a more national criterion. As we have suggested, the genetic turn in current times especially fits nation-state countries. The idea of a productionist "people's-home country" as promoted by the Swedish Social Democrats or Zionists in Israel is enticing for liberal countries that want to protect the hegemony of their larger ethnic national community.

The second strategy is the path that the Nordic welfare countries we have studied in this book have silently taken. The question we have analyzed thus is whether these countries could preserve their welfare system and be productive. John Glad in his seminal work of the future of human evolution, wonders what the state should do vis-à-vis welfare systems that care for the weakest members of society setting a pattern of waste that will harm society in the future "... What to do? Deny poor women and their children financial assistance? Bribe the upper classes into childbearing? Or throw up our hands in dismay and allow society to be genetically dumbed-down?" (Glad 2006: 56).

Precisely what we have shown in this book is that which John Glad defines as "bribing the upper classes," is precisely what welfare societies are doing. Eugenics in this sense is not used in its negative form related to sterilization but in a positive way, tending to increase the reproductive capacity of the national stock.

There is no question that some moral questions should be raised regarding this move. That is one of the main reasons why the whole rationale of these practices is concealed under a veil of progressive universal realm. ART and family allowances and leaves, especially destined for lesbians and homosexuals, are a progressive universal policy. However, at the same time, it is quite clear that the sectors of the population these policies are directed at are the productive modern nationals of the country. These policies, thus, are obviously not relevant for traditional immigrant communities.

In sum, European countries aspire to preserve their national culture, foster to keep from being welfare societies, and are eager to be competitive in the world economy. It seems that these new policies are part of a non-publicly declared idea. Finally, we claim that although moral criticism cannot be dismissed, we suggest that there is no evil in the fact that liberal majorities strive to preserve a national liberal societal culture. This can be done either by assimilation of the newcomers, or by increasing fertility of the modern national stock, as we have seen in the countries studied, or both tracks completing each other.

Notes

Introduction

1 Economic Eugenics? http://economix.blogs.nytimes.com/2008/09/28/economic-eugenics/?pagemode.

2 The primary feature of eugenics was the conviction that human traits of character, for good or evil, are genetically transmitted.

3 In the United States for much of the first half of the twentieth century, the mentally ill, the retarded, alcoholics, recent immigrants, and those thought to be sexually promiscuous, especially if they were members of minority groups and poor, became the object of government-sponsored sterilization efforts aimed at preventing the spread of "bad" genes to future generations (Proctor 1988).

4 While some immigrant groups in Europe, such as Hindus and East Asians, enjoy relatively low unemployment rates and healthy incomes, the largest immigrant group, Muslims, has become a burden for the European welfare systems. This has led to extensive cutbacks in public services in order to keep up with welfare payments—closing clinics and emergency rooms, reducing staff in hospitals, cutting police and military spending, eliminating course offerings at public universities, and so on.

5 There was a debate during the 1980s in Europe regarding the meaning of the concept of free movement of persons. While some member states understood that this should apply to the European Union, others argued in favor of free movement for everyone, which would mean an end to internal border checks altogether. The final decision, due to the fact that there is no agreement, was that France, Germany, Belgium, Luxembourg, and the Netherlands decided in 1985 to create a territory without internal borders. This became known as the Schengen area.

6 In this, Europe remains lastingly different from the classic immigrant nations, in which from the very outset the legal immigrant is considered a fully functioning and rightful member of the new society.

7 To view the introduction of biopolitics to the bioethics discourse, see Bishop and Jotterand (2006), Garrard and Wilkinson (2003), Jasanoff (2005), Lysaught (2009), Pellegrino (2000), Tremain (2005, 2006, 2008).

8 Gunner Broberg and Mattias Tydén have pointed out how Swedish anthropology contributed to the investigation of Nordic head shapes, height, and weight, but by the end of the nineteenth century it was already being succeeded by the study of biological inheritance in humans, animals, and plants.

1 Eugenics and productive socialism

1 By modeling the notion that history and institutions count, ideas emerged to explain some welfare policies' realities. In cross-national comparisons, ideas have been invoked to explain not that ideas matter but how and when they mattered. Following Marshall

and Rose (1989), ideas are defined as the solution to collective action, involve complementary configurations of individuals, and institutions such as nation state, class, or religion that motivate individuals to actions they would not take had they not been grouped under the ideational abstraction. For ideational studies in public policy, see Berman (1998: Introduction), Hall (1993), Hansen (2000), McNamara (1998).

2 For extensive literature on American eugenics see Cooke (1998), Kevles (1985), Kline (2001), Larson (1995), Ludmerer (1972), Pickens (1968), Stepan (1991).

3 Drawing from the literature on the history of social sciences and technology studies, it is argued that the rise and development of modern ideologies was shaped by the rise of modern science, by the growing prestige of technology and new modes of production, and by the development of publics whose favorable judgment of modern science was rooted in the decline of older authoritative modes of thought. Thus, in Chapter 1 we propose to depart from over-simplistic identification of eugenics, as sociobiological thought that never achieved ideological "takeoff" in terms of influence, and offer an alternative mode of explanation.

4 For a general conceptual discussion of race, see Banton (1977, 1998).

5 Buffon, Blummenbach, Gregoire, and others did not attach social hierarchies to racial differences. They stressed the unity within human diversity. See excerpts from Buffon, Blumenbach, and Gregoire in Chukwudi Eze (1997).

6 According to Bowler (1989b), anthropologists and archaeologists of the mid-nineteenth century extended the human scale of development to include biological as well as social evolution within the idea of progress.

7 Recent historical studies analyze race conceptualizations during the nineteenth century by using an analogy between racial and cultural categories, such as gender, to develop the nineteenth-century set of racial meanings, which diverge from those of the eighteenth century. See in particular, Sciebinger (1989) and Stocking (1968).

8 For the historical relations between genetics and eugenics, see Bowler (1989a, 1989b), Ludmerer (1972), and MacKenzie (1976).

9 Haldane (1932: 18–47) outlined the problematic nature of Pearson's theory of genetics: Mendelian justification for the theory of selection in genetics.

10 Pearson's statistical data were designed to substantiate his political policy and practice. See MacKenzie (1979).

11 Recent works have drawn attention to the inconsistency of the theory of evolution to explain the various modes of degeneration in thought during the nineteenth century. These works present manifold manifestations of degeneration in sociology, politics, demography, literature, and fiction. See Bowler (1984), Chamberlin and Gilman (1985), Greenslade (1994), Pick (1989).

12 For the histories of the idea of progress during the nineteenth century, see for instance Bury (1987) and Nisbet (1980).

13 Galton (1869) discussed the degeneration of the Ancient Greeks.

14 Karl Pearson, *National Life from the Standpoint of Science*, cited by Semmel (1960: 41).

15 We have based our analysis on highly selected material. This methodological choice can be defended with regard to the virtually limitless number of ideological writings of the twentieth century that are relevant to present research.

16 Sternhell has pointed out that fascist ideology resulted from the convergence of an integralist and conservative right and an anti-Marxian socialist left. Both revolutionary syndicalist followers of Georges Sorel and national integralist followers of Charles Maurras come together in *Cahiers du Cercle Prohudon*, in 1910. They found a common ground in their attack on the Third Republic and on the theoretical creation of a new type of national socialism, the basis of the national syndicalist state.

17 Alfredo Rocco, who became the draftsman of Mussolini's key judicial and labor legislation of 1925–1927, even before World War I was a strong advocate of the *marcia dei produttori*, the organization of producer groups in a national and authoritarian corporative structure. See Maier (1987).

18 Mussolini's myth of the nation was central to the creation of a fascist doctrine on the synthesis of nationalism with socialism.
19 On "youth ideology" as phrased by Eisenstadt to denote the moral significance of youth for the creation of a national identity, see Eisenstadt (1956: 311).
20 In the era of growing nationality and patriotism, the frequency of death and sickness in the youthful male population became an agenda for an ongoing discourse concerning the relationship between state policy and medical responsibility. For such a critical analysis, see Houghton (1985), Linton (1991), and Mosse (1985).
21 In a speech concerning Germany's foreign policy to more than 6000 students delivered in Munich on January 26, 1936, Hitler said: "The white race is destined to rule. This is its unconscious urge which arises from an heroic conception of life and which is entirely non-pacifist" (1969: 1258).
22 The important point, however, is that the regime's statutes called for the establishment of genetic health courts and supreme genetic health courts in order to determine the racial health of the population.
23 To see the progressive side of eugenics, see Freeden (1979).
24 Of other countries that implemented similar laws, we might mention the United States, Great Britain, and Japan. Both the proposals for the "feeble minded" under the Mental Deficiency Act of 1913 and sexual sterilization of mental defectives in the 1930s in Great Britain stand against our views of a "liberal" Britain in the early twentieth century. Towards the end of the nineteenth century, the European theory of degeneration began to also gain acknowledgment in the United States. The first explicit eugenics legislation was enacted in Connecticut in 1896. Indiana, Washington, and especially California followed this line. In 1913 the state of California was empowered to sterilize any inmate of any institution diagnosed as suffering from hereditary insanity.
25 A popular song in the 1920s expressed these views. Words by Gus Kahn and Raymond B. Egan. Music by Richard A Whiting; copyright Warner Bros Inc., 1921.
26 Wells' *Sociological Papers, Proceedings of the Sociological Society*, London, 1902. Cited in Ray (1983).
27 Eden Paul, *Socialism and Eugenics*, Manchester: National Labour Press, 1911. Cited in Paul (1998: 13).
28 In his book *Imperialism, the Highest Stage of Capitalism*, written in the summer of 1916, Lenin drew the conclusion that capitalism had entered the highest stage of its development—imperialism. "Capitalism has grown into a world system of colonial oppression and of financial strangulation of the overwhelming majority of the population of the world by a handful of 'advanced' countries" (Lenin 1960, vol. I: 715).
29 Geertz's model of text known as "thick description" shares with Bourdieu a commitment to interpretive sociology. The interpretive account of ideology has sought to retain the core elements that are central to any reflexive study. Bourdieu views science and ideology as distinctive fields of cultural production. Indeed, the structural characterization of an ideology as a network of relations between different positions determines the form of struggles internal to that field and varies socially and historically. See Bourdieu (1990), Bourdieu and Wacquant (1992), Geertz (1973: 204).

2 From "race hygiene" to "national-productivist hygiene"

1 Zaremba (August 20–21, 1997). Zaremba's articles received international attention especially when republished by the British *Independent* on August 30, 1997.
2 Norway might also be portrayed as a country where eugenics and sterilization were tools of social reform. However, Norway is not included among the countries that expose a high level of permissive ART programs in current times, and in Chapter 2 we deal with examples that exemplify the shift from welfare national productivist regimes to permissive ART programs.

3 We should stress that although Sweden adopted multicultural policies in the 1960s, the principle of a people's-home productivist society still remains. In more senses than one, Swedish multiculturalism and Danish assimilationist policies are not so distant from each other. For a different view see Tawat (2011).

4 The wounds sustained in the Civil War were alleviated by conciliatory measures such as including the Social Democrats in the government; in 1926–1927 they formed a minority government on their own.

5 Gunner Broberg and Mattias Tydén have pointed out how Swedish anthropology contributed to the investigation of Nordic head shapes, height, and weight; but by the end of the nineteenth century this was already being succeeded by the study of biological inheritance in humans, animals, and plants.

6 Petrén held that sterilization of the mentally deficient was a necessary policy instrument of life-long institutionalization. The mentally deficient should be sterilized if it could be proved that they had no chance of recovery. Accordingly, the epileptic should be allowed sterilization on humanitarian grounds as it was the only possibility for these otherwise healthy individuals to marry.

7 Axel Danielsson and Hjalmar Branting, who wrote the Swedish Social Democratic program in 1897, believed that social democracy was socially revolutionary in the sense that it demanded a very thorough change of society. However, they believed that should be done through universal suffrage. See Jenson and Sineau (2001), Richard and Mahon (2001).

8 See, for example, Broberg and Roll-Hansen (1996: 108–109). This is the most recent and most thorough study of eugenics and welfare politics.

9 See the controversy highlighted by the articles on the subject written by journalist Maciej Zaremba that appeared in the liberal Swedish daily *Dagens Nyheter*, August 20–21, 1997. The articles were published as part of a broader attack on the social welfare state, the so-called "Swedish folk-home."

10 It should be noted that in recent studies there is no common agreement on the significance of the racial aspect of Ruppin's works. See, for example, the views of Sergio DellaPergola (1999), Arthur Goren (2005), Mitchell Hart (2000), Amos Morris-Reich (2006), and Derek Penslar (1991).

11 The Mandatory Government controlled Categories A, B, and D, but granted partial authority to the Zionist Organization for Category C in exchange for a promise by the movement to guarantee the maintenance of such immigrants in their first year of residence. See Halamish (2006: 61).

12 The estimated calculation is based on the Central Bureau of Statistics, retrieved from www1.cbs.gov.il/reader/?MIval=%2Fprices_db%2FMachshevon_Results.html&MySubject=11120010&Subjects=38&MyCode=11120010&MultMin=19511016&MultMax=20110111&DateMin=16%2F10%2F1951&DateMax=11%2F01%2F2011&SpecificationType=1&ssum=23000&koeff=10000&Days_2=11&Months_2=1&Years_2=2011&Days_1=16&Months_1=10&Years_1=1951.

13 For continued concern about institutionalization in 1946, see General Council of the Jewish Community of Palestine (1946: 44–46).

14 Nordau recommended that Jews live in nature and pursue a more physical culture—that Judaism build some muscles. See Presner (2003). See also Biale (1992: 178–179), Falk (1998: 594), Nordau (1936: 171–172, 174, 176; 1968), Soder (1991). For sports and Zionism, see Kaufman (2005).

15 From being one of poorest countries in Europe in the 1920s, Sweden became one of the richest in the late 1960s. Only Japan and the former West Germany come close to Sweden in this respect.

16 There were other options to sterilization. One of them was controlled family care (*kontrolleret familiepleje*), where patients were put under the custody of a family outside the institution; however, under the limited supervision of an institution. This provision was comparable to the provision of guardianship in the 1913 Mental Deficiency Act.

17 Even Catholics, although reluctant to accept that women should have the right to suspend reproduction, still favor positive eugenics for the cause of population growth.
18 To be precise, these later ideas of national culture were advanced and promoted first and foremost by the Radical Liberal Party, especially by the author and priest Grundtvig, who stressed the idea of Danishness in the aftermath of the defeat of 1864.
19 Still, in more practical ways the differences are not so huge. Both promote a productivist integration, which implies the assumption of cultural values. The lack of integration of immigrants into the productive norms of Sweden in the later years is frequently interpreted as a source of social problems. It is also clear that Denmark has a more determined idea of assimilation into the nation.
20 Virtanen (1985: 514) argues that the Finnish-Swede race was academically categorized in terms of a hierarchy by French physical anthropologist Count Arthur de Gobineau around 1853. He divided hierarchically between "whites," "yellows," and "blacks," and argued, based on physical evidence, that Finns were "yellow," something promoted by the supposedly "white" Finnish-Swedes to further their authority to rule a supposedly inferior race.
21 In the 1920s, the features of an East Baltic race were first identified with the Finnish-speaking people; however, as research progressed it was discovered that some characteristics of that race were also present among the Swedish-speaking population.
22 One of the arguments explaining the zeal to become European is the fact that it backs the current position in the European Union, which Finland joined in 1995.
23 After the defeat, the Reds created their own social and cultural organizations and stayed away from the bourgeois organizations. However, with time, the worldview of the Whites became dominant in public service.
24 This new trend is related to the new spirit expanded in the 1960s when the first non-eugenic laws of sterilization were passed in the world. In all Scandinavian countries the right to contraceptive sterilization was introduced to all citizens from the age of 25. The 1985 Sterilization Law in Finland defined sterilization as a family planning method. See Ekman and Liljestrand (1993).

3 The bioethical turn to biopolitics: the emergence of bio-nationalism

1 To view the introduction of biopolitics to the bioethics discourse, see Bishop and Jotterand (2006), Garrard and Wilkinson (2003), Jasanoff (2005), Lysaught (2009), Pellegrino (2000), Tremain (2005, 2006, 2008).
2 Michael Foucault, "Les mailles du pouvoir," cited in Curtis (2002).
3 Following Farrell (2005: 105), Foucault used the terms biopower and biopolitics interchangeably: "Foucault describes the technologies used to manage populations as *'biopolitics' or 'biopower'*" [emphasis added]. See also the same approach addressed by Ojakangas (2005) and Parry (2005).
4 Although it was Foucault (1979[1997]) who popularized the concept of biopolitics, earlier discussions on biopolitics in modernity were attempted by Harris (1911) and Roberts (1938). For recent discussions, see Somit (1972). For more recent developments, see Carmen (1997), Dickinson (2004), Simons (2006).
5 For a recent series of related works on governing populations by exclusionary practices, see Alatout (2006), Amoore (2006), Gledhill (2005), Legg (2005), Philo (2001), Sparke (2006).
6 Foucault exemplified the dividing practice of discourse in the case of Pierre Rivière, who became the victim of exclusionary psychiatric discourse through the production of sanity and insanity divisions. See Foucault (1975a).
7 In contrast to Foucault's division between sovereign and disciplinary forms of rule (1975b[1991]), Hardt and Negri (2001) synthesize such power techniques into a total sovereign power. Following Hardt and Negri: "The absoluteness of imperial power is

the complementary term to its complete immanence to the ontological machine of production and reproduction, and thus to the biopolitical context" (2001: 41) and "Rule is exercised directly over the movements of productive and cooperating subjectivities" (2001: 319).

8 In March 2005 the European Commission launched an open debate on demographic change with "Confronting Demographic Change: a New Solidarity between the Generations ('Green Paper')." The Commission presented its conclusions from this debate in the 2006 follow-up communication, "The Demographic Future of Europe—From Challenge to Opportunity ('the Communication')." Much relevant statistical and other information can be found in "Europe's Demographic Future: Facts and Figures on Challenges and Opportunities" of October 2007 ("Facts and Figures"). Finally, the European Parliament adopted a resolution on the demographic future of Europe on February 21, 2008—"The Resolution."

9 To help inform EU policy deliberations, analysts from RAND Europe and RAND U.S. examined the relationships between policy and demographic change. The RAND team analyzed the interrelationships between European government policies and demographic trends and behavior, and assessed which policies could prevent or mitigate the adverse consequences of current low fertility and population aging. The monograph *Low Fertility and Population Ageing: Causes, Consequences, and Policy Option* documents the study's findings.

10 Administration for Children and Families—U.S. Department of Health. Retrieved from www.acf.hhs.gov/programs/cb/systems/ncands/ncands98/glossary/glossary.htm.

11 The cost of the procedure itself is comparatively easy to calculate, but the cost of the infertility work-up, pregnancy, and neonatal care, and the cost for ART-associated morbidity and mortality are difficult to assess.

12 When including births following insemination, more than 6 percent of all children born in Denmark are conceived after some form of assisted reproduction.

13 Dansk Fertilitetsselskab (Danish Fertility Society) report on ART in Denmark. Retrieved from www.ertilitetsselskab.dk/Aarsrapport.htm.

4 The bio-regulation return of eugenics for a new century of massive immigration and multiculturalism

1 In cross-national comparison, ideas have been invoked to explain not that ideas matter but how and when they mattered. Following Marshall and Rose (1989), ideas are defined as the solution to collective action, involve complementary configurations of individuals and institutions such as nation state, class, or religion that motivate individuals to actions they would not take had they not been grouped under the ideational abstraction. The ideational features of path dependency have been delineated as factors influencing welfarist policy and politics.

2 Weiss (2000) shows that international or supra-national organizations such as the EU treat employment and family as separate spheres, and argues for a strong progressive family model that "corresponds to social reality." The reluctance of the EU to interconnect family planning and productivist social policies could be exemplified in the way the EU initially gave priority to heterosexual and legally married couples in social security benefits. The European Court has moderated discrimination based on family status and promoted gender equity. The court has required payment of parental benefits to both mothers and fathers, and part-time work to be given hourly wages equal to that of full-time work, but has excluded unpaid family work from social security benefits.

3 Due to its oil revenues, Norway escaped most of the fiscal difficulties that hit its Nordic neighbors. This did not, however, stop restructuring from occurring, due to economic problems similar to those in the other Nordic countries (Marklund and Nordlund 1999).

4 Retrieved from http://stats.oecd.org/Index.aspx?DatasetCode=CSP2010.

5 The German poverty atlas shows a huge social divide: 18.5.2009. Retrieved from www.dw-world.de/dw/article/0,,4262772,00.html.
6 Interview with Sergio Della Pergola, July 6, 2008.
7 Sweden continues to have one of the most generous and welcoming policies in Europe for immigrants and refugees; however, there is a growing tendency towards restrictions on immigration policy. Indeed, recent studies have demonstrated that Sweden's massive immigration questions Sweden's multiculturalism policy (Dahlstedt and Hertzberg 2007).
8 A recent report stated that immigration costs Sweden at least 40–50 billion Swedish kröner every year, probably several hundred billion, and has greatly contributed to bringing the Swedish welfare state to the brink of bankruptcy. Immigration costs in Sweden amount to almost 30 percent of the public budget following Jasson's book (2002), which showed an increase in immigration costs. I am not sure how his book relates to the increase in immigration costs.
9 Swedish Government, Written Government Communication 2001/02:129: 37.
10 National Health Insurance Law art. 12 § 5.2 (1995). Retrieved from www.un.org/esa/gopher-data/ga/cedaw/17/country/israel/C-ISR-P3.EN.
11
    1 In appeal no. 998/96, *Dr. Tal Yerus-Hakak v. The Ministry of Health*, the appellant argued that the Health Ministry's regulation is discriminatory on the basis of personal status and sexual inclination. Moreover, the appellant argued that the government regulation is invalid because it violates the basic human right to be a parent, over his body and his privacy; because it exceeds the Ministry's authority; because it is in secondary and not primary legislation; because it is unreasonable; and because it was not published.
    2 A single-parent family is defined as a family where one parent runs a household for him/herself and his/her children (to age 17), and who does not have a permanent partner (based on Svirsky *et al.* 2002).

12 When Social Democrats returned to power in 1994, they had to reform the welfare state and reduce benefits in a somewhat unbalanced system, which led to big losses to the left in 1998; these losses were mostly regained in 2002 when the Social Democrats won and continue in government backed by the Left and Green parties.

5 European countries and eugenic productivism gospel—potential convergence trends in eugenic productivist population policy

1 *Lov om integration af udlændinge i Danmark*. LOV nr. 474 af 01.07. 1998. 11 (Ejrnæs 2001).
2 The cost of the procedure itself is comparatively easy to calculate, but the cost of the infertility work-up, pregnancy, and neonatal care, and the cost for ART-associated morbidity and mortality is difficult to assess.
3 Dansk Fertilitetsselskab (Danish Fertility Society) report on ART in Denmark. Retrieved from www. ertilitetsselskab.dk/Aarsrapport.htm.
4 When including births following insemination, more than 6 percent of all children born in Denmark are conceived after some form of assisted reproduction.
5 Surrogacy is practiced, and all forms of donor conception are available in Finland (Hazekamp and Hamberger 2005).

Conclusions

1 On post-national citizenship, see Soysal (1997). On the stability of nationalism, see Brubaker (1992). In Europe, see also Al Sayyad and Castells (2002), Nielsen (1999), Shadid and Van Koningsveld (1995).

Bibliography

Adams, M.B. (ed.) (1990) *The wellborn science: Eugenics in Germany, France, Brazil, and Russia*. New York: Oxford University Press.

Agamben, G. (1998) *Homo sacer: sovereign power and bare life*. Stanford: Stanford University Press.

Agar, N. (1995) "Designing babies: Morally permissible ways to modify the human genome," *Bioethics*, 9: 1–15.

Agence de la biomédicine (2007) *Le guide de l'assistance medicale a la procreation*. Saint-Denis La Plaine Cedex, France: Agence de la biomédicine. Online. Available HTTP: www.procreationmedicale.fr/dl/2008/02/guideamp-bd.pdf.

Al Sayyad, N. and Castells, M. (2002) *Muslim Europe or Euro-Islam: Politics, culture, and citizenship in the age of globalization*. Lanham, MD: Lexington Books.

Alatout, S. (2006) "Towards a bio-territorial conception of power: Territory, population, and environmental narratives in Palestine and Israel," *Political Geography*, 25: 601–621.

Albæk, E. (2003) "Political ethics and public policy: Homosexuals between moral dilemmas and political considerations in Danish parliamentary debates," *Scandinavian Political Studies*, 26(3): 245–267.

Alesina, A., Glaeser, E. and Sacerdote, B. (2001) 'Why doesn't the United States have a European-style welfare state?" *Brookings Papers on Economic Activity*, 2.

Allen, G.E. (1989) "Eugenics and American social history, 1880–1950," *Genome*, 31(2): 885–889.

Allen, G.E. (2001) "Is a new eugenics afoot?" *Science*, 294(5540): 59–61.

Amoore, L. (2006) "Biometric borders: Governing mobilities in the war on terror," *Political Geography*, 25: 336–351.

Andersen, T.M. (2004) "Challenges to the Scandinavian Welfare Model," *European Journal of Political Economy*, 20: 743–754.

Anderson, B. (1986) *Imagined communities: Reflections on the origin and spread of nationalism*. London: Verso.

Andersson, G. (2000) "The impact of labor-force participation on childbearing behavior: Pro cyclical fertility in Sweden during the 1980s and 1990s," *European Journal of Population*, 16: 293–333.

Anttonen, A. and Sipalä, J. (1996) "European social care services: Is it possible to identify models?" *Journal of European Social Policy*, 6(2): 87–100.

Arendt, H. (1943) "We refugees." Originally published in the *Menorah Journal*. Reprinted in R.H. Feldman (ed.) *The Jew as pariah: Jewish identity and politics in the modern age* (pp. 55–67). New York: Grove Press.

Aristotle (1885) *Politics*. Book One. Part II; trans. B. Jowett. Oxford: Clarendon Press.

Bahle, T. and Rothenbacher, F. (1998) "Germany," in J. Ditch, H. Barnes, and J. Bradshaw (eds.) *Developments in national policies in 1996* (pp. 37–61). Brussels: European Commission.

Baker, M. and Tippin, D. (1999) *Poverty, social assistance and the employability of mothers: Restructuring welfare states*. Toronto: University of Toronto Press.

Balibar, E. (1991) *Race, nation, class: ambiguous identities*. London, New York: Verso.

Ball, S. (1988) "The moral aspects of socialism (1896)," in S. Trombley (ed.) *The right to reproduce: A history of coercive sterilization* (pp. 290–322). London: Weidenfeld and Nicolson.

Banton, M. (1977) *The idea of race*. London: Tavistock Publications.

Banton, M. (1998) *Racial theories* (2nd edn.). Cambridge: Cambridge University Press.

Barkan, E. (1992) *The retreat of scientific racism*. Cambridge: Cambridge University.

Barrios, S., Görg, H., and Strobl, E. (2003) "Multinational enterprises and new trade theory: Evidence for the convergence hypothesis," *Open Economies Review*, 14(4): 397–418.

Barzilai, G. (2003) *Communities and law: Politics and cultures of legal identities*. Ann Arbor: University of Michigan Press.

Bauer, T., Larsen, C., and Matthiessen, P.C. (2004) "Immigration policy and Danish and German immigration," in T. Tranæs and K.F. Zimmermann (eds.) *Migrants, work and the welfare state* (pp. 31–73). Odense: University Press of Southern Denmark and the Rockwool Foundation Research Unit.

Baynes, N.H. (ed.) (1969) *The Speeches of Adolf Hitler (April 1922–August 1939)*. New York: Howard Ferting.

Baynton, D. (2005) "Defectives in the land: Disability and American immigration policy 1882–1924," *Journal of American Ethnic History*, 24(3): 31–44.

Benda Report (1985) *In-vitro-Fertiliation, Genomanalyse und Gentherapie. Bericht der gemeinsamen Arbeitsgruppe des Bundesministers für Forschung und Technologie und des Bundesministers der Justiz.* Hrsg. vom Bundesminister für Forschung und Technologie, Bonn, Redaktionelle Bearbeitung Peter Lange. München Schweitzer.

Benhabib, S. (2002) *The claims of culture: Equality and diversity in the global era*. Princeton, NJ: Princeton University Press.

Bennett, J.H. (ed.) (1983) *Natural selection, heredity and eugenics*. Oxford: Oxford University Press.

Bergmark, A. and Palme, J. (2003) "Welfare and unemployment crisis: Sweden in the 1990s," *International Journal of Social Welfare*, 12: 108–122.

Berman, S. (1998) *The social democratic moment: Ideas and politics in the making of interwar Europe*. Cambridge, MA: Harvard University Press.

Berman, S. (2006) *The primacy of politics: Social democracy and the making of Europe's twentieth century*. Cambridge: Cambridge University Press.

Bernhardt, E. (2002) "Cohabitation and marriage among young adults in Sweden: Attitudes, expectations and plans. Nordic demography: trends and differentials," *Scandinavian Population Studies*, 13: 157–170.

Bertelsmann Foundation (2003) *International reform monitor, 7*. Gutersloh, Germany. Online. Available HTTP: www.reformmonitor.org.

Biale, D. (1992) *Eros and the Jews*. New York: Basic Books.

Bishop, J.P. and Jotterand, F. (2006) "Bioethics as biopolitics," *Journal of Medicine and Philosophy*, 31: 205–212.

Bleiklie, I., Goggin, M., and Rothmayr, C. (eds.) (2004a) *Governing assisted reproductive technology: A cross-country comparison*. London: Routledge.

Bleiklie, I., Goggin, M., and Rothmayr, C. (eds.) (2004b) *Comparative biomedical policy: Governing assisted reproductive technologies*. London: Routledge.

Blume, K., Gustafsson, B. Pedersen, P.J., and Verner, M. (2007) "At the lower end of the table: Determinants of poverty among immigrants to Denmark and Sweden," *Journal of Ethnic and Migration Studies*, 33(3): 373–396.

Boas, F. (1916) "New evidence in regard to the instability of human types," *Proceedings of the National Academy of Sciences*, 2(12): 713–718.

Boas, F. (1928) *Anthropology and modern life*. New York: W.W. Norton and Company.

Boas, F. (1938) *The mind of primitive man* (rev. edn.). New York: Greenwood Press (Original work published 1911).

Boas, F. (1940) *Race, language and culture*. New York: The Free Press.

Bock, G. (1986) *Zwangssterilisation im nationalsozialismus: Studien zur rassenpolitik und frauenpolitik*. Opladen: WDV.

Bourdieu, P. (1990) *In other words*. Stanford: Stanford University Press.

Bourdieu, P. and Wacquant, L.J.D. (1992) *An invitation to reflexive sociology*. Chicago: University of Chicago Press.

Bowler, P.J. (1984) *Evolution: The history of an idea*. Berkeley: University of California.

Bowler, P.J. (1989a) *The Mendelian revolution*. London: The Athlone Press.

Bowler, P.J. (1989b) *The invention of progress*. Oxford: Basil Blackwell.

Box, J.F. (1978) *R.A. Fisher: The life of a scientist*. New York: Wiley.

Bradley, D. (1999) "Comparative family law and the political process: Regulation of sexual morality in Finland," *Journal of Law & Society*, 26(2): 175.

Bradshaw, J. and Finch, N. (2002) *A comparison of child benefit packages in 22 countries*. UK: Department for Work and Pensions. Online. Available HTTP: www.dwp.gov.uk/asd/asd5/rrep174.asp.

Broberg, G. and Roll-Hansen, N. (eds.) (1996) *Eugenics and the welfare state: Sterilization policy in Denmark, Sweden, Norway and Finland*. Michigan: Michigan State University Press.

Broberg, G. and Tydén, M. (1996) "Eugenics in Sweden: Efficient care," in G. Broberg and N. Roll-Hanson (eds.) *Eugenics and the welfare state: Sterilisation policy in Denmark, Sweden, Norway and Finland*. East Lansing, MI: Michigan State University Press.

Brubaker, R. (1992) *Citizenship and nationhood in France and Germany*. Cambridge, MA: Harvard University Press.

Bryld, M. (2001) "The infertility clinic and the birth of the lesbian. The political debate in assisted reproduction in Denmark," *European Journal of Women's Studies*, 8(3): 299–312.

Buchanan, A., Brock, D.W., Daniels, N., and Wikler, D. (2000) *From chance to choice: Genetics and justice*. Cambridge: Cambridge University Press.

Budge, I. and Robertson, D. (1987) "Do parties differ, and how? Comparative discriminant and factor analyses," in I. Budge, D. Robertson, and D. Hearl (eds.) *Ideology, strategy and party change: Spatial analysis of post-war elections programs in 19 democracies* (pp. 387–416). Cambridge: Cambridge University Press.

Bujard, M., Dorbritz, J., Grünheid, E. *et al.* (2012) *(Keine) Lust auf Kinder? Geburtenentwicklung in Deutschland*. Bundesinstitut für Bevölkerungsforschung, Online. Available HTTP: www.bibdemografie.de/SharedDocs/Publikationen/DE/Download/Broschueren/keine_lust_auf_kinder_2012.pdf?__blob=publicationFile&v=9.

Bunton, R. and Petersen, A. (eds.) (2005) *Genetic governance: Health, risk and ethics in the biotech era*. London and New York: Routledge.

Buruma, I. (2010) "Europe turns Right," in *Project Syndicate: A world of ideas*. Online. Available HTTP: www.project-syndicate.org/commentary/buruma42/English.

Bury, J.B. (1987) *The idea of progress*. New York: Dover Publications.

Cadman, L. (2009) "Life and death decisions in our posthuman(ist) times," *Antipode*, 41(1): 133–158.

Caplan, A., McGee, G., and Magnus, D. (1999) "What's immoral about eugenics?" *British Medical Journal*, 319: 1284–1285.

Carlson, E.A. (2001) *The unfit: A history of a bad idea*. Woodbury, NY: Cold Spring Harbor Laboratory Press.

Carmen, I.H. (1997) "Biopolitics: The newest synthesis?" *Genetica*, 99(2/3): 173–184.

Castles, F.G. (2004) *The future of the welfare state: Crisis myths and crisis realities*. Oxford: Oxford University Press.

Castles, F.G. and Pierson, C. (1995) "New convergence? Recent policy developments in the United Kingdom, Australia and New Zealand," *Policy and Politics*, 24(3): 233–245.

Castles, S. and Miller, M. (1993) *The age of migration*. London: Macmillan.

Centers for Disease Control and Prevention (1998) "Implementation of the Fertility Clinic Success Rate and Certification Act of 1992; proposed model program for the certification of embryo laboratories," *Federal Register*, 63(215): 60178.

Central Bureau of Statistics (2004) *Live births, deaths, natural increase, infant deaths and stillbirths, by population group*. Online. Available HTTP: www.cbs.gov.il/reader/shnaton/templ_shnaton_e.html?num_tab=st03_02&CYear=2010.

Central Bureau of Statistics (2007–2008) *Households and families demographic characteristics*. Online. Available HTTP: www.cbs.gov.il/webpub/pub/text_page_eng.html?publ=46&CYear=2008&CMonth=1.

Central Bureau of Statistics (2010) *Society in Israel*. Online. Available HTTP: www.cbs.gov.il/webpub/pub/text_page.html?publ=54&CYear=2008&CMonth=1.

Chamberlin, J.A. and Gilman, S.L. (eds.) (1985) *Degeneration: The dark side of progress*. New York: Columbia University Press.

Chambers, G.M., Chapman, M.G., Grayson, N., Shanahan, M. and Sullivan, E.A. (2007) "Babies born after ART treatment cost more than non-ART babies: A cost analysis of inpatient birth-admission costs of singleton and multiple gestation pregnancies," *Human Reproduction*, 22: 3108–3115.

Christopher, K. (2002) "Welfare state regimes and mothers' poverty," *Social Politics*, Spring: 60–66.

Chukwudi Eze, E. (ed.) (1997) *Race, and the enlightenment: A reader*. Cambridge: Blackwell.

Cole, M. (1946) *Beatrice Webb*. New York: Harcourt, Brace and Company.

Cooke, K.J. (1998) "The limits of heredity: Nature and nurture in American eugenics before 1915," *Journal of the History of Biology*, 31: 263–278.

Corrêa, S. and Reichman, R. (1994) *Population and reproductive rights: Feminist perspectives from the south*. London: Zed Press.

Council of Europe (2003) *Recent demographic developments in Europe, 2003*. Strasbourg: Council of Europe Publishing.

Craig, G., Dietrich, H., and Gautié, J. (2005) "Excluded youth or young citizens? Ethnicity, young people and the labour market in three EU countries," in H. Bradley and J. v. Hoof (eds.) *Young people in Europe: Labour markets and citizenship*. Bristol: The Polity Press.

Crampton, J.W. (2007) "The biopolitical justification for geosurveillance," *Geographical Review*, 97(3): 389–403.

Curtis, B. (2002) "Foucault on govermentality and population: The impossible discovery," *Canadian Journal of Sociology*, 27: 505–533.

Dahl, R. (1989) *Democracy and its critics*. New Haven, CT: Yale University Press.

Dahlstedt, M. and Hertzberg, F. (2007) "Democracy the Swedish way? The exclusion of 'immigrants' in Swedish politics," *Scandinavian Political Studies*, 30(2): 175–203.

Dahrendorf, R. (1995) *A history of the London School of Economics 1895–1995*. Oxford: Oxford University Press.

Damgaard, E. (2004) "Developments in Danish parliamentary democracy: Accountability, parties and external constraints," *Scandinavian Political Studies*, 27(2): 115–131.

Danielsson, A. (1972) *Om revolutionen i Sverige*. Stockholm: Arbetarkultur.

Darwin, C. (1874) *The descent of man and selection in relation to sex*. New York: A.L. Burt.

Davenport, C.B. (1910) *Eugenics*. New York: Eugenics Record Office at Cold Spring Harbor; New York: Henry Holt.

Davidovitch, N. and Shvarts, S. (2005) "Health and Zionist ideology: Medical selection of Jewish European immigrants to Palestine," in I. Borowy and W.D. Gruner (eds.) *Facing illness in troubled times: Health in Europe in the interwar years, 1918–1939*. Frankfurt: Peter Lang Publishing.

DellaPergola, S. (1999) "Arthur Ruppin revisited: The Jews of today, 1904–1994," in S.M. Cohen and G. Horencyzk (eds.) *National variations in Jewish identity: Implications for Jewish education* (pp. 53–84). New York: State University of New York Press.

DellaPergola, S. and Even, J. (eds.) (2001) *Papers in Jewish demography*. Jerusalem: Hebrew University.

DESTATIS (2007) Press Release No. 481/2007–11–28. "Share of alternative family forms higher in eastern than in western Germany."

Dickinson, E.R. (2004) "Biopolitics, fascism, democracy: Some reflections on our discourse about 'modernity'," *Central European History*, 37(1): 3–4.

Dobbelsteen, S.H.A.M., Gustafsson, S.S., and Wetzels, C.M.M.P. (2000) *Child care in the Netherlands between government, firms and parents: Is the deadweight loss smaller than in the public daycare system of Sweden?* Amsterdam: University of Amsterdam.

Doron, A. and Kramer, R.M. (1991) *The welfare state in Israel*. Boulder, CO: Westview.

Dowbiggin, I. (2002) "A rational coalition: Euthanasia, eugenics, and birth control in America, 1940–1970," *Journal of Policy History*, 14: 223–260.

Duelund, P. (2003) "Cultural policy in Denmark," in P. Duelund (ed.) *The Nordic cultural model: Nordic cultural policy in transition*. Copenhagen: Nordic Cultural Institute.

Duster, T. (2003) *Backdoor to eugenics*. London: Routledge.

Dutton, E. (2008) "Battling to be 'European': Myth and the Finnish race debate," *Antrocom*, 4(2): 171–179.

Ehrlich, J. (December 19, 2003) "Liberal Netherlands grows less so on immigration," *The Christian Science Monitor*. Online. Available HTTP: www.hartford-hwp.com/archives/61/315.html.

Eisenstadt, S.M. (1956) *From generation to generation*. Glencoe: Free Press.

Eitan, C. (1937) "Machalot atzabim bein yeladim vetotzaoteihen hasocialiot," *Bulletin on Social Welfare in Palestine*, 7–8: 122–123.

Ejrnæs, M. (2001) "Integrationsloven—en case, der illustrerer etniske minoriteters usikre medborgerstatus," *AMID Working Paper Series*. No. 1.

Ekerwald, H. (2001) "The modernist manifesto of Alva and Gunnar Myrdal: Modernization of Sweden in the thirties and the question of sterilization," *International Journal of Politics, Culture and Society*, 14(3): 539–561.

Ekman, B. and Liljestrand, J. (1993) "Experiences of legal sterilization in Scandinavia," *Contraception*, 43: 157–158.

Elkeles, T. and Seifert, W. (1996) "Immigrants and health: Unemployment and health-risks of labour migrants in the Federal Republic of Germany 1984–1992," *Social Science & Medicine*, 43(7): 1035–1047.

Ellis, H.E. (1912) *The Task of Social Hygiene*. London: Constable & Company Ltd.

Engbersen, G. (2003) "Spheres of integration: Towards a differentiated and reflexive ethnic minority policy," in R. Sackmann, B. Peters, and T. Faist (eds.) *Identity and integration: Migrants in Western Europe* (pp. 59–76). Aldershot/Burlington: Ashgate.

Engeli, I. (2009) "The challenges of abortion and assisted reproductive technologies policies in Europe," *Comparative European Politics*, 7: 56–74.

Erler, W. and Erler, D. (2007) "Germany," in P. Moss and K. Wall (eds.) *International review of leave policies and related research 2007* (pp. 157–172). London: Department for Business, Enterprise and Regulatory Reform.

ESHRE (2001) The European IVF-monitoring programme (EIM), for the European Society of Human Reproduction and Embryology (ESHRE). "Assisted reproductive technology in Europe, 1997. Results generated from European registers by ESHRE," *Human Reproduction*, 16: 384–391.

ESHRE (2002) The European IVF-monitoring programme (EIM), for the European Society of Human Reproduction and Embryology (ESHRE). "Assisted reproductive technology in Europe, 1999. Results generated from European registers by ESHRE," *Human Reproduction*, 17: 3260–3274.

ESHRE (2006) The European IVF monitoring programme (EIM) for the European Society of Human Reproduction and Embryology (ESHRE). "Assisted reproductive technology in Europe, 2002. Results generated from European registers by ESHRE," *Human Reproduction*, 21: 1680–1697.

ESHRE (2008) *Focus on reproduction*. Grimbergen, Belgium: ESHRE.

Esping-Andersen, G. (1985) *Politics against markets*. Princeton, NJ: Princeton University Press.

Esping-Andersen, G. (1990) *The three worlds of welfare capitalism*. Cambridge: Polity Press.

Esping-Andersen, G. (1999) *Social foundations of post-industrial economies*. Oxford: Oxford University Press.

Èstergaard-Nielsen, E. (2003) "Debates and developments theme issue: The rise of the far right in urban Europe counting the cost: Denmark's changing migration policies," *International Journal of Urban and Regional Research*, 27(2): 448–454.

European Commission (2007a) *Report on equality between women and men—2007*. Online. Available HTTP: http://ec.europa.eu/social/main.jsp?catId=738&langId=en&pubId=17&type=2&furtherPubs=yes.

European Commission (2007b) *Social protection in the Member States of the European Union, of the European Economic Area and in Switzerland. Organisation of social protection—charts and descriptions*. Brussels, Belgium: MISSOC.

Evans, S. (2010) "Germany's charged immigration debate," *BBC News Europe*, October 17, 2010.

Expatica (2004) "Tracking the Dutch immigration policy." Online. Available HTTP: www.expatica.com/nl/essentials_moving_to/essentials/tracking-the-dutch-immigration-policy-7692_9566.html.

Fair, L.S. (2008) "The new diversity in Denmark: Integration challenges for Danes, immigrants and refugees," *International Journal of Diversity in Organisations, Communities & Nations*, 8(5): 225–237.

Falk, R. (1998) "Zionism and the biology of the Jews," *Science in Context*, 11: 587–607.

Fallaize, E.N. (1925) "The study of primitive races with special reference to forms of marriage," *The Eugenics Review*, 17: 77–87.

Farrell, C.O. (2005) *Michel Foucault*. London: Sage Publications.

Federal Ministry for Family Affairs, Senior Citizens, Women and Youth (2004) *Families in Germany—facts and figures*. Online. Available HTTP: www.iesf.es/fot/Families-Germany-Facts-Figures-2004.pdf.

Federal Ministry for Family Affairs, Senior Citizens, Women and Youth (2007) *Overview of the standard benefits and tax concessions relevant to family policy and the amounts involved in the Federal Republic of Germany*. Online. Available HTTP: www.bmfsfj.de/bmfsfj/generator/RedaktionBMFSFJ/Abteilung2/Pdf-Anlagen/regelleistungen-steuererm_C3_A4_C3_9Figungen-englisch,property=pdf,bereich=,sprache=de.rwb=true.pdf.

Finnish Ministry of Social Affairs and Health (1999) *Finnish family policy. Helsinki*. Online. Available HTTP: www.vn.fi/stm/english/index.htm.

Fisher, I. (1921) "Impending problems of eugenics," *Scientific Monthly*, 13(3): 214–231.

Fisher, R.A. (1930) *The genetical theory of natural selection*. Oxford: Clarendon Press.

Fitzpatrick, T. (2001) "Before the cradle: New genetics, biopolicy and regulated eugenics," *Journal of Social Policy*, 30: 589–612.

Flora, P. and Heidenheimer, A.J. (eds.) (1981) *The development of the welfare states in Europe and America*. New Brunswick, NJ: Transaction Books.

Fonnest, I. de la F., Søndergaard, F., Fonnest, G., and Vedsted-Jacobsen, A. (2000) "Attitudes among health care professionals on the ethics of assisted reproductive technologies and legal abortion," *Acta Obstetricia et Gynecologica Scandinavica*, 79: 49–53.

Foucault, M. (1970) *The order of things: An archaeology of human sciences*. New York: Vintage Books.

Foucault, M. (1972a) *The archaeology of knowledge*, trans. A.M. Sheridan-Smith. London: Tavistock Publications.

Foucault, M. (1972b) "Discourse on language." In *The archeology of knowledge and the discourse on language*, trans. A.M. Sheridan Smith. New York: Pantheon Books.

Foucault, M. (1975a) *I, Pierre Rivière, having slaughtered my mother, my sister and my brother: A case of parricide in the nineteenth century*. London: University of Nebraska Press.

Foucault, M. (1975b[1991]) *Discipline and punish. The birth of the prison*, trans. A. Sheridan. London: Penguin Books.

Foucault, M. (1976[2003]) *Society must be defended*, trans. D. Macey. London: Penguin.

Foucault, M. (1979[1997]) "The birth of biopolitics," in P. Rabinow (ed.) *Michel Foucault, ethics: Subjectivity and truth* (pp. 73–79). New York: The New Press.

Foucault, M. (1980) "Two lectures," in C. Gordon (ed.) *Power/knowledge: Selected interviews and other writings* (pp. 78–108). New York: Pantheon.

Foucault, M. (1982) "Afterword: Subject and power," in H.L. Dreyfus and P. Rabinow (eds.) *Michel Foucault: Beyond structuralism and hermeneutics*. Chicago: University of Chicago Press.

Foucault, M. (1988) *Politics, philosophy, culture: Interviews and other writings 1977–1984*, ed. L. Kritzman. New York: Routledge.

Foucault, M. (1989a) *The history of sexuality: An introduction*, trans. R. Hurley. New York: Vintage Books.

Foucault, M. (1989b) *Foucault live: Collected interviews, 1961–1984*. New York: Semiotext(e).

Foucault, M. (2007) "Spaces of security: The example of the town. Lecture of 11 January 1978," in M. Senellart (ed.) *Security, territory, population* (pp. 1–27), trans. G. Burchell. Lectures at the Collège de France, 1977–1978, [2004] 2007. Basingstoke, England: Palgrave Macmillan.

Fraser, N. (2001) "Recognition without ethics?" *Theory, Culture & Society*, 18: 21–42.

Freeden, M. (1979) "Eugenics and progressive thought: A study in ideological affinity," *The Historical Journal*, 22(3): 645–679.

Freeden, M. (1983) "Eugenics and ideology," *The Historical Journal*, 26(4): 959–962.

Friedlander, D. and Goldscheider, C. (1979) *The population of Israel*. New York: Columbia University Press.

Furedy, F. (1997) *Population and development*. Cambridge: Polity Press.

Galal, L.P (2006) "Medborgerskab og kulturel mangfoldighed" [Citizenship and cultural diversity], *UFE-nyt*, special issue: 8–18.

Galton, F. (1869) *Heredity genius: An inquiry into laws and consequences*. London: Macmillan.

Galton, F. (1883) *Inquiries into human faculty*. London: Macmillan.

Galton, F. (1904) "Eugenics: Its definition, scope and aims," *Nature*, 70(1804): 82.

Garrard, E. and Wilkinson, S. (2003) "Does bioethics need moral theory?" in M. Hayry and T. Takala (eds.) *Scratching the surface of bioethics* (pp. 35–45). Amsterdam, The Netherlands: Rodopi.

Geddes, A. (2003) *The politics of migration and immigration in Europe*. London: Sage Publications.

Geertz, C. (1973) *The interpretation of cultures*. New York: Basic Books.

Gems, D. (1999) "Politically correct eugenics," *Theoretical Medicine and Bioethics*, 20(2): 201–213.

General Council of the Jewish Community of Palestine (1946) *Social welfare service of the Jewish community*. Jerusalem.

Glad, J. (2006) *Future human evolution: Eugenics in the twenty-first century*. Schuylkill Haven, PA: Hermitage Publishers.

Gledhill, J. (2005) "The power of ethnic nationalism: Foucault's bio-power and the development of ethnic nationalism in Eastern Europe," *National Identities*, 7(4): 347–368.

Goethe Institute (September 1, 2010) Online. Available HTTP: www.goethe.de/kug/buw/sub/ein/en24571.htm.

Goldberg, D.T. (1997) *Racial subjects: Writing on race in America*. New York: Routledge.

Goren, A.Y. (2005) *Arthur Ruppin, his life and work*. Jerusalem: Yad Tabenkin [Hebrew].

Gowricharn, R. (2002) "Integration and social cohesion: The case of the Netherlands," *Journal of Ethnic and Migration Studies*, 28(2): 259–273.

Gradstein, M. and Schiff, M. (2006) "The political economy of social exclusion, with implications for immigration policy," *Journal of Population Economics*, 19(2): 327–344.

Graig, G. (2002) "Ethnicity, racism and the labour market: A European perspective," in J. Goul Andersen and P. Jensen (eds.) *Changing labour markets, welfare policies and citizenship* (pp. 149–183). Bristol: The Polity Press.

Graig, G., Dietrich, H., and Gautié, J. (2005) "Excluded youth or young citizens? Ethnicity, young people and the labour markets in three EU countries," in H. Bradley and J. van Hoof (eds.) *Young people in Europe: Labour markets and citizenship*. Bristol: The Policy Press.

Granberg, M., Nygren, K.G., and Wikland, M. (2005) "The cost of IVF in the Nordic countries," *Human Reproduction*, 20(Suppl 1): i100.

Grant, J., Hoorens, S., Gallo, F., and Cave, J. (2006) "Should ART be part of a population policy mix? A preliminary assessment of the demographic and economic impact of assisted reproductive technologies," *The RAND Corporation*. Online. Available HTTP: www.rand.org.

Green-Pedersen, C. (1999) "The Danish welfare state under bourgeois reign: The dilemma of popular entrenchment and economic constraints," *Scandinavian Political Studies*, 22: 243–260.

Greenslade, W. (1994) *Degeneration, culture and the novel 1880–1940*. Cambridge: Cambridge University Press.

Gregor, J. (1979) *The young Mussolini and the intellectual origins of fascism*. Berkeley, CA: University of California Press.

Gross, R. and Brammli-Greenberg, S. (2000) *The health and welfare of women in Israel: Findings from a national survey* (RR-361–00). Jerusalem: JDC–Brookdale Institute.

Gross, R. and Brammli-Greenberg, S. (2001) "The experiences of women and men in the health system: Findings of a national survey," paper presented at The Scientific Conference on Women's Health 2002, organized by The National Council for Women's Health, Jerusalem.

Guillén, M. (2001) "Is globalization civilizing, destructive or feeble? A critique of five key debates in the social science literature," *Annual Review of Sociology*, 27: 235–260.

Gustafsson, B., Johansson, M., and Palmer, E. (2003) "Economic well-being of children in the deep Swedish recession of the 1990s," in J. Bradshaw (ed.) *Children and social security* (pp. 37–62). Aldershot: Ashgate.

Haave, P. (2000) *Sterilizering av tatere 1934–1977. En historisk undersøkelse av lov og praksis* [Sterilization of vagrants 1934–1977. A historical analysis of legislation and practice]. Oslo: Research Council of Norway.

Haave, P. (2001) "Tvangssterilizering i Norge" [Enforced sterilization in Norway], "Frihed, lighed og velfærd" [Freedom, equality and welfare], report to the 24th Nordic Conference of Historians, vol. 2: 140–162.

Haddon, A.C. (1909) *The races of man and their distribution*. Cambridge: Miler & Company. (Reprinted by Cambridge University Press, 2012.)

Häkli, J. (1999) "Cultures of demarcation: Territory and national identity in Finland," in G.H. Herb (ed.) *Identities, nationalism, territory and scale*. Lanham, MD: Rowman & Littlefield.

Halamish, A. (2006) "A new look at immigration of Jews from Yemen to Mandatory Palestine," *Israel Studies*, 11(1): 59–78.

Haldane, J.B.S. (1932) *The causes of evolution*. London, New York and Toronto: Longmans, Green and Co.

Hall, P.A. (1993) "Policy paradigms, social learning and the state: The case of economic policy making in Britain," *Comparative Politics*, 25(3): 275–296.

Halpern, L. (1937) "Insanity among the Jews in Palestine," *Harefuah*, 12: 203–219.

Halpern, L. (1938) "Machalot haruach bein hayehudim beEretz Israel," Special publication in *Harefuah*, 12(4): 6.

Halvorsen, R. and Jensen, P.H. (2004) "Activation in Scandinavian welfare policy: Denmark and Norway in a comparative perspective," *European Societies*, 6: 461–483.

Hansen, B.S. (1996) "Something rotten in the state of Denmark," in G. Broberg and N. Roll-Hansen (eds.) *Eugenics and the welfare state*. East Lansing, MI: Michigan State University Press.

Hansen, R. (2000) *Citizenship and immigration in postwar Britain: The institutional origins of a multicultural nation*. Oxford: Oxford University Press.

Hansen, R. (2009) "The poverty of post-nationalism: Citizenship, immigration and the new Europe," *Theory and Society*, 38: 1–24.

Hansen, R. and King, D. (2001) "Eugenic ideas, political interests, and policy variance: Immigration and sterilization policy in Britain and the U.S.," *World Politics*, 53: 237–263.

Hardt, M. and Negri, A. (2001) *Empire*. Boston, MA: Harvard University Press.

Harris, G.W. (1911) "Biopolitics," *The New Age*, 10(9): 197.

Hart, M.B. (2000) *Social science and the politics of modern Jewish identity*. Stanford, CA: California University Press.

Hartmann, B. (1987) *Reproductive rights and wrongs: The global politics of population control and reproductive choice*. New York: Harper and Row.

Hartmann, B. (1997) "Population control in the new world order," in D. Eade (ed.) *Development for health, development in practice reader* (pp. 80–85). Oxford: Oxfam.

Haub, C. (2007) *2007 World population data sheet*. Washington, DC: Population Reference Bureau.

Häyrinen-Alestalo, M. and Kallerud, E. (2004) "Towards a biotech society—Nordic perspectives," in M. Häyrinen-Alestalo and E. Kallerud (eds.) *Mediating public concern in biotechnology. A map of sites, actors and issues in Denmark, Finland, Norway and Sweden* (pp. 7–22). Oslo: NIFU Report 1/2004.

Hazekamp, J. and Hamberger, L. (2005) "Regulation of artificial reproductive technology: The Nordic experience," in P.R. Brinsden (ed.) *Textbook of in-vitro fertilization and assisted reproduction*. 3rd edn. (pp. 435–441). London: Taylor & Francis.

Heclo, H. (1974) *Modern social politics in Britain and Sweden: From relief to income maintenance*. New Haven: Yale University Press.

Hedetoft, U. (2009) "Denmark versus multiculturalism," in S. Vertovec and S. Wessendorf (eds.) *Multicultural backlash: European discourses, policies and practices*. Florence, KY: Routledge.

Heidenheimer, A.J., Heclo, H., and Adams, C.T. (1990) *Comparative public policy: The politics of social choice in Europe and America*. 3rd edn. New York: St. Martin's Press.

Hemminki, E., Rasimus, A., and Forssas, E. (1997) "Sterilization in Finland: From eugenics to contraception," *Social Science Medicine*, 45(12): 1875–1884.

Henderson, C. (1900) "Science in philanthropy," *Atlantic Monthly*, 85(508): 249–254.

Herskovits, M.J. (1973) *Frans Boas, augustism*. Clifton, NJ: Kelley Publishers.

Hietala, M. (1985) "In search of the Finnish female type," in Aira Kemiläinen (ed.) *Mongols or Germanics?* Helsinki: Finnish Historical Society.

Hietala, M. (1996) "From race hygiene to sterilization: The eugenic movement in Finland," in G. Broberg and N. Roll-Hansen (eds.) *Eugenics and the welfare state: Sterilization policy in Denmark, Sweden, Norway and Finland*. Michigan: Michigan State University Press.

Hitler, A. (1924) *Mein kampf*, trans. R. Manheim. Boston: Houghton Mifflin Company. Reprinted 1971.

Hitler, A. (1928) *Hitler's secret book*, trans. S. Attanasio. New York: Grove Press. Reprinted 1961.

Hitler, A. (1969) *The speeches of Adolf Hitler*, April 1922–August 1939, Vol. I, ed. N.H. Baynes. New York: Howard Fetig.

Hobsbawm, E. (1964) *Labouring men: Studies in the history of labour*. London: Weidenfeld and Nicholson.

Hoem, B. and Hoem, J. M. (1996) "Sweden's family policies and roller-coaster fertility," *Journal of Population Problems*, 52(3–4): 1–22.

Hoem, J. (2008) "The impact of public policies on European fertility," *Demographic Research*, 19(10): 249–260. Online. Available HTTP: www.demographic-research.org/Volumes/Vol. 19/10/19–10.pdf.

Holm, L.K. (2006) "Folketinget og udlændingepolitikken—Diskurser om naturaliserede, indvandrere og flygtninge 1973–2002" (Ph.D.-afhandling, Aalborg: Akademiet for migrationstudier i Danmark, Aalborg universitet).

Holzinger, K., Knill, C., and Sommerer, T. (2011) "Is there convergence of national environmental policies? An analysis of policy outputs in 24 OECD countries," *Environmental Politics*, 20(1): 20–41.

Honkasalo, B. (1934) "Sterilissoinista," *Sielunterveyssuran Aikakaullehti*, 7(7): s8.

Houghton, W. (1985) *The Victorian frame of mind*. New Haven and London: Yale University Press.

Huxley, J., Haddon, A.C., and Carr-Saunders, A.M. (1935) *We Europeans*. London: Penguin Books.

ICMART (2009a) "International Committee for Monitoring Assisted Reproductive Technology (ICMART) and the World Health Organization (WHO), revised glossary of ART terminology, 2009," *Fertilization and Sterilization*, 92: 1520–1524.

ICMART (2009b) "International Committee for Monitoring Assisted Reproductive Technology (ICMART). World collaborative report on in vitro fertilization, 2002," *Human Reproduction*, 1: 1–11.

Immervoll, H., Sutherland, H., and De Vos, K. (2000) "Reducing child poverty in the European Union: The role of child benefits," in K. Vleminckx and T. Sneeding (eds.) *Child well-being, child poverty and child policy in modern nations* (pp. 407–432). Bristol: The Polity Press.

INSEE. Government of France. (2008) "Tabc3—Tableau complémentaire 3: Taux de fécondité par groupe d'âges." Online. Available HTTP: www.insee.fr/fr/ffc/ipweb/ip1220/ip1220.xls [French].

Jahn, D. (2006) "Globalization as Galton's problem: The missing link in the analysis of diffusion patterns in welfare state development," *International Organization* 60(2): 401–431.

Jamison, A. (2002) "From innovation to appropriation: On the politics of technoscience," in *Focus on biotechnology: Issues related to R&D in biotechnology: Denmark in a comparative perspective*. The Danish Institute for Studies in Research and Research Policy.

Jasanoff, S. (2005) *Designs on nature: Science and democracy in Europe and the United States*. Princeton, NJ: Princeton University Press.

Jasson, L. (2002) *Mångfall eller Välfärd*. Kategori: Ekonomi Bokörlaget Vitsippan.

Jeandidier, B. and Albiser, E. (2001) *To what extent do family policy and social assistance transfers equitably reduce the intensity of child poverty? A comparison between the US, France, Great Britain and Luxembourg*. Luxembourg Income Study Working Paper No. 255.

Jenson, J. and Sineau, M. (eds.) (2001) *Who cares? Women's work, childcare, and welfare state redesign*. Toronto: University of Toronto Press.

Jönsson, I. (2003) "Fertility changes and family policy in Sweden," in M.-T. Letablier and S. Pennec (eds.) *Changing family structure in Europe: New challenges for public policy*. Loughborough: European Research Centre.

Joppke, C. (2004) "The retreat of multiculturalism in the liberal state: Theory and policy," *British Journal of Sociology*, 55(2): 237–257.

Joppke, C. (2007) "Transformation of immigrant integration civic integration and antidiscrimination in the Netherlands, France, and Germany," *World Politics*, 59(2): 243–273.

Joppke, C. (2010) "Identity," in *Citizenship and immigration* (pp. 111–144). Cambridge: Polity Press.

Judt, T. (2003) "Israel: The alternative," *New York Review of Books*, October 23, 2003.

Kamerman, S.B. and Kahn, A.J. (1997) "Investing in children: Government expenditures for children in western industrialized societies, 1960–1990," in S. Danziger and G.A. Cornia (eds.) *Child poverty in industrialized societies* (pp. 91–121). Oxford: Oxford University Press.

Kamerman, S.B. and Kahn, A.J. (2000) "Child and family policies in an era of social policy retrenchment and restructuring," in K. Vleminckx and T. Smeeding (eds.) *Child well-being in modern nations* (pp. 501–526). Bristol: The Polity Press.

Kanaaneh, R.A. (2002) *Birthing the nation: Strategies of Palestinian women in Israel*. Berkeley: University of California Press.

Kartovaara, L. and Sauli, H. (2001) *Children in Finland*. Helsinki: Statistics Finland, Population 2001.

Katz, S. (July 30, 2003) "Pro-lifers: Hadash's abortion bill a weapon in 'demographic war'," *Jerusalem Post*, p. 3.

Katznelson, B. (1944) *Ktvim*. Tel Aviv: Israeli labor party (Mapai).

Kaufman, H. (2005) "Jewish sports in the Diaspora, Yishuv and Israel: Between nationalism and politics," *Israel Studies*, 10(2): 147–167.

Kautto, M., Heikkilä, M., Hvinden, B., Marklund, S., and Ploug N. (1999) "Introduction: The Nordic welfare states in the 1990s," in M. Keikkila, B. Hvinden, M. Kautto, S. Marklund, and N. Ploug (eds.) *Nordic social policy* (pp. 1–18). London: Routledge.

Kautto, M., Fritzell, J., Hvinden, B., Kvist, J., and Uusitalo, H. (2001) "Conclusion: Nordic welfare states in the European context," in M. Kautto, J. Fritzell, B. Uvnden, J. Kvist, and H. Uusitalo (eds.) *Nordic welfare states in the European context* (pp. 262–273). London: Routledge.

Keilman, N. (2008) "Concerns in the European Union about low birth rates," *European Review*, 7: 333–340.

Kelsey, J. (1995) *Economic fundamentalism*. London: Pluto Press.

Kemiläinen, A. (1998) *Finns in the shadow of the Aryans: Race theories and racism*. Helsinki: Finnish Historical Society.

Kemp, T. (June 28, 1933) "Sterilisation og Racehygiene," *Politiken*.

Kemp, T. (1957) "Genetics hygiene experiences in Denmark in recent years," *The Eugenics Review*, 49: 18.

Kent, M. and Haub, C. (2005) "Global demographic divide," *Population Bulletin*, 60(4): 1–23.

Kenworthy, L. (1999) "Do social welfare policies reduce poverty? A cross-national assessment," *Social Forces*, 77: 1119–1139.

Kerr, A. and Shakespeare, T. (2002) *Genetic politics: From eugenics to genome*. Cheltenham, UK: New Clarion Press.

Kerr, C. (1983) *The future of industrial societies: Convergence or continuing diversity?* Cambridge, MA: Harvard University Press.

Kevles, D.J. (1985) *In the name of eugenics: Genetics and the uses of human heredity*. Berkeley: University of California Press.

Kevles, D.J. (1992) "Out of eugenics: The historical politics of the human genome," in D.J. Kevles and L. Hood (eds.) *The code of codes: Scientific and social issues in the human genome project* (pp. 3–36). Cambridge, MA/London: Harvard University Press.

Kim, L. (February 1, 2002) "Germany tightens stem-cell imports," *Christian Science Monitor*. Online. Available HTTP: www.csmonitor.com/2002/0201/p08s01-woeu.html.

Kimmerling, B. and Migdal, J.S. (2002) *Palestinians: The making of a people* (enlarged edn.). New York: Free Press.

King, D. (1999) *In the name of liberalism*. Oxford: Oxford University Press.

King, D. (2000) *Making Americans: Immigration, race, and the origins of a diverse democracy*. Cambridge, MA: Harvard University Press.

King, D. and Rueda, D. (2008) "Cheap labor," *Perspectives on Politics*, 6(2): 279–297.

Kinsella, K. and Velkoff, V.A. (2001) *An aging world: 2001*. Washington, DC: U.S. Census Bureau, Series P95/01–1.

Kirejczyk, M. (1999) "Parliamentary cultures and human embryos: The Dutch and British debates compared," *Social Studies of Science*, 29(6): 889–912.

Kline, W. (2001) *Building a better race: Gender, sexuality and eugenics from the turn of the century to the baby boom*. Berkeley: University of California Press.

Knox, R. (1850) *The races of men: A fragment*. London: Henry Renshaw.

Koch, L. (2000) *Tvangssterilization i Danmark 1929–67* [Enforced sterilization in Denmark 1929–67]. Copenhagen: Gyldendal.

Koch, T. (2003) "Genetics, eugenics, and discrimination: At the beginning of the end of the Dark Ages of medicine," *Gemellogical Review*, 3: 47–56.

Kofman, E. (2002) "Contemporary European migrations, civic stratification and citizenship," *Political Geography*, 28(1): 1035–1054.

Kohler, H.-P. (2000) "Social interactions and fluctuations in birth rates," *Population Studies*, 54(2): 223–237.

Koopman, R. (2010) "Trade-offs between equality and difference: Immigrant integration, multiculturalism and the welfare state in cross-national perspective," *Journal of Ethnic and Migration Studies*, 36: 1–26.

Kop, Y. (ed.) (2003) *Israel's social services 2003*. The Herbert M. Singer annual report series annual socio-economic report November 20, 2003. Jerusalem: The Israel Center for Social Policy Studies.

Korpi, W. (1983) *The democratic class struggle*. London: Routledge and Kegan Paul.

Korpi, W. (2003) "Welfare-state regress in Western Europe: Politics, institutions, globalization, and Europeanization," *Annual Review of Sociology*, 29: 589–609.

Korpi, W. and Palme, J. (1998) "The paradox of redistribution and the strategy of equality: Welfare state institutions, inequality and poverty in the Western countries," *American Sociological Review*, 63(5): 661–687.

Korpi, W. and Palme, J. (2003) "New politics and class politics in the context of austerity and globalization: Welfare state regress in 18 countries, 1975–1995," *American Political Science Review*, 97: 425–446.

Kosonen, P. (2001) "Globalization and the Nordic welfare states," in M.R. Sykes, B. Palier, and P. Prior (eds.) *Globalization and European welfare states: Challenges and change* (pp. 153–172). Basingstoke: Palgrave.

Kramnick, I. and Sheerman, B. (1993) *Harold Laski: A life on the left*. New York: Allen Lane Penguin.

Kristensen, C. (2007) "Towards two-tier universalism? A discussion of the welfare principle," *Journal of Social Sciences*, 3(2): 52–59.

Kulawik, T. (2004) "The politics of stem cells in Sweden: Explaining liberal regulations in the social democratic state," paper presented at the EASST Conference, Paris, August 2004.

Kvist, R. (1994) "The racist legacy in modern Swedish Saami policy," *Canadian Journal of Native Studies*, 14(2): 203–220.

Larner, W. (1997) "The legacy of the social: Market governance and the consumer," *Economy and Society*, 26: 373–399.

Larson, E.J. (1995) *Sex, race and science: Eugenics in the deep south*. Baltimore: John Hopkins University Press.

Larsson, T. (2003) "Swedish cultural policy," in P. Duelund (ed.) *The Nordic cultural model: Nordic cultural policy in transition*. Copenhagen: Nordic Cultural Institute.

Lawler, P. (2007) "Janus-faced solidarity: Danish internationalism reconsidered," *Cooperation and Conflict*, 42(1): 101–126.

Legg, S. (2005) "Foucault's population geographies: Classifications, biopolitics and governmental spaces," *Population, Space and Place*, 11(3): 137–156.

Leira, A. (2002) *Working parents and the welfare state: Family change and policy reform in Scandinavia*. Cambridge: Cambridge University Press.

Lemke, T. (2002) "Genetic testing, eugenics and risk," *Critical Public Health*, 12: 283–290.

Lenin V.I. (1916[1939]) *Imperialism, the highest state of capitalism: A popular outline*. New York: International Publishers.

Lenin, V.I. (1960) *Selected works*. Vol. I and II. Moscow: Gospolitizdat.

Lenoir, N. (1999) "Universal declaration on the human genome and human rights," *The Columbia Human Rights Law Review*, 30(1): 537–587.

Letablier, M.-T. and Pennec, S. (eds.) (2003) *Changing family structure in Europe: New challenges for public policy*. Loughborough, UK: European Research Centre.

Lewis, R. (2005) *Finland: Cultural lone wolf*. New York: Intercultural Press.

Lindqvist, S. (August 30, 1997) "Valfard stoppade steriliseringar," *Dagens Nyheter*.

Linton, D.S. (1991) *"Who has the youth, has the future": The campaign to save young workers in Imperial Germany*. New York: Cambridge University Press.

Lombardo, P.A. (2008) *Three generations, no imbeciles: Eugenics, The Supreme Court, and* Buck V. Bell. Baltimore: Johns Hopkins University Press.

Lotringer, S. (ed.) (1989) *Foucault live* (Collected Interviews, 1961–1984), trans. L. Hochroth and J. Johnston. New York: Semiotext(e).

Ludmerer, K.M. (1972) *Genetics and American society*. Baltimore and London: The Johns Hopkins University Press.

Lüttichau, I. (2004) "'We are family': The regulation of 'female-only' reproduction," *Social & Legal Studies*, 13(1): 81–101.

Lynn, R. (2001) *Eugenics: A reassessment*. Santa Barbara, CA: Prager.

Lynöe, N. (2007) "Race enhancement through sterilization: Swedish experiences," *International Journal of Mental Health*, 36(1): 17–25.

Lysaught, M. (2009) "Docile bodies: Transnational research ethics as biopolitics," *Journal of Medicine and Philosophy*, 34(4): 384–408.

MacKenzie, D. (1976) "Eugenics in Britain," *Social Studies of Sciences*, 6: 499–532.

MacKenzie, D. (1979) "Karl Pearson and the professional middle class," *Annals of Science*, 36: 125–143.

Mahoney, J. (2000) "Path dependence in historical sociology," *Theory and Society*, 29(4): 507–548.

Maier, C. (1987) "The economics of fascism and Nazism." In *In search of stability: Explorations in historical political economy*. Cambridge: Cambridge University Press.

Malin, M. and Burrell, R. (2004) "Finland: Unregulated practices, familiarity and legality," in E. Blyth and R. Landau (eds.) *Third party assisted conception across cultures: Social, legal and ethical perspectives* (pp. 73–93). London and New York: Jessica Kingsley Publishers.

Malmberg-Heimonen, I. and Julkunen, I. (2006) "Out of unemployment? A comparative analysis of the risks and opportunities longer-term unemployed immigrant youth face when entering the labour market," *Journal of Youth Studies*, 9(5): 575–592.

Mammey, U. and Schwarz, K. (2002) "The demographic characteristics of immigrant populations in Germany," in W. Haug, P. Compton, and Y. Courbage (eds.) *Demographic characteristics of immigrant populations. Population studies* no. 38 (pp. 193–244). Strasbourg: Council of Europe.

Mannion, G. (2006) "Genetics and the ethics of community," *Heythrop Journal*, XLVII: 226–256.

Marklund, S. and Nordlund, A. (1999) "Economic problems, welfare convergence and political instability," in M. Heikkila, B. Hvinden, M. Kautto, S. Marklund, and N. Ploug (eds.) *Nordic social policy* (pp. 19–53). London: Routledge.

Marsden, R. (1999) *The nature of capital: Marx after Foucault*. London: Routledge.

Marshall, A. (1897) "The old generation of economists and the new," *Quarterly Journal of Economics*, 11(2): 115–135.

Marshall, G. and Rose, D. (1989) *Social class in modern Britain*. London: Unwin Hyman.

Marshall, T.H. (1950) *Citizenship and social class, and other essays*. Cambridge: Cambridge University Press.

Martin, C. (1999) *France*. Family Observer. Luxembourg: Office for Official Publications of the European Communities.

Martin, P.L. (1998) *Germany: Reluctant land of immigration, AICGS German Issues*, Monograph 21. Washington, DC: American Institute for Contemporary German Studies.

Matmon, A. (1933) *Improving the human race and its value to our nation*. Tel-Aviv: Biological-Hygienic Library.

Mayr, E. (1976) *Evolution and the diversity of life*. Cambridge: Harvard University Press.

Mayr, E. (1991) *One long argument: Charles Darwin and the genesis of modern evolutionary thought*. Cambridge: Harvard University Press.

Mazumdar, P.M.H. (1992) *Eugenics, human genetics and human failings*. London: Routlege.

McNamara, K. (1998) *The currency of ideas: Monetary politics in the European Union*. Ithaca, NY: Cornell University Press.

Mead, W.R. (1993) *An experience of Finland*. London: C. Hurst and Co.

Mei-Ami, N. (2005) "Background paper on sperm donation in Israel," paper presented to the Committee on the Rights of the Child, Research and Information Centre, The Knesset, Jerusalem, March 2005.

Meir, Y. (1952) "Increasing birth rates or increasing fertility rates?" *Eytanim*, 3–4: 76–77.

Migdal, J. (2001) *Through the lens of Israel*. New York: State of New York University Press.

Miles, R. (1989) *Racism*. London, New York: Routledge.

Mills, E.M. (1933) *Census of Palestine 1931*, Vol. I. Palestine Part I Report (pp. 224–241). Alexandria: Whitehead Morris Limited;.

Mishra, R. (1984) *The welfare state in crisis*. Brighton, UK: Wheatsheaf Books.

Mishra, R. (1990) *The three worlds of welfare capitalism*. Cambridge: Polity Press.

Montpetit, E., Rothmayr, C., and Varone, F. (eds.) (2007) *The politics of biotechnology in North America and Europe: Policy networks, institutions and internationalization*. Lanham, MD: Lexington Books.

Morris-Reich, A. (2006) "Arthur Ruppin's concept of race," *Israel Studies*, 11(3): 1–30.

Morrissens, A. and Sainsbury, D. (2005) "Migrants' social rights, ethnicity and welfare regimes," *Journal of Social Policy*, 34: 637–660.

Mosse, G.L. (1985) *Nationalism and sexuality: Respectability and abnormal sexuality in modern Europe*. New York: H. Fetig.

Mouritsen, P. (2006) "The particular universalism of a Nordic civic nation: Common values, state religion, and Islam in Danish political culture," in T. Modood, R. Zapata-Barrero, and A. Triandafyllidou (eds.) *Multiculturalism, Muslims and citizenship* (pp. 70–93). London: Routledge.

Mouritsen, P. (2009) "The resilience of citizenship traditions: Civic integration in Germany, Great Britain, and Denmark," article presented to audiences at the closing conference in Berlin September 24–25 2009, of the EMILIE project—A European Approach to multicultural citizenship: Legal, political and educational challenges—funded by the European Commission Research DG, 6th Framework Program; at the "Beyond Identity Politics" conference, Stockholm University, October 21–22, 2009; and at Aarhus University's Francis Fukuayama seminar, October 26, 2009.

Murphy, C. (May 6, 2002) "France chastened by election crisis," *BBC News Online*.

Mussolini, B. (1914) "The war as a revolutionary event," in R. Griffin (ed.) *Fascism* (pp. 26–27). Oxford: Oxford University Press. Reprinted 1995.

Mussolini, B. (1917a) "Fascism's myth: The nation," in R. Griffin (ed.) *Fascism* (pp. 43–44). Oxford: Oxford University Press. Reprinted 1995.

Mussolini, B. (1917b) "Trenchocracy," in R. Griffin (ed.) *Fascism*. Oxford: Oxford University Press. Reprinted 1995.

Mussolini, B. (1929) "The achievements of the fascist revolution," in R. Griffin (ed.) *Fascism* (pp. 62–65). Oxford: Oxford University Press. Reprinted 1995.

Mussolini, B. (1933) *The corporate state* (2nd edn.). Florence: Vallecchi Publisher.

Mussolini, B. (1935) *The doctrine of fascism*. Firenze: Vallecchi Editore.

Myers-JDC-Brookdale Institute (2000) *Use of health services survey 1999/2000*. Jerusalem: Myers-JDC-Brookdale Institute.

Myrdal, A. (1968) *Nation and family: The Swedish experiment in democratic family and population policy*. Cambridge, MA: MIT Press.

Myrdal, A. and Myrdal, G. (1935) *Kris i befolkningsfragan* [Crisis in the population question] (3rd revised and extended edn.). Stockholm: Albert Bonniers forlag.

Myrdal, G. (1938) "Population problems and policies," *The Annals*, 197: 203–204.

Nachman, S. (1929) "Race, people and nationalism," in *Nachman Syrkin's writings* (pp. 262–267). Collected by B. Katzenelson and Y. Kufman. Tel-Aviv: Davar Publications [Hebrew].

Nielsen, J.S. (1999) *Towards a European Islam*. Basingstoke/New York: Palgrave.

Nisbet, R.A. (1980) *History of the idea of progress*. New York: Basic Books.

Nordau, M. (1936) "Yehadut hashririm," in *Max Nordau el 'amo: ketavim mediniim*. Tel-Aviv: Medinit Press.

Nordau, M. (1968) *Degeneration*. Lincoln: University of Nebraska Press.

Nordlund, A. (2000) "Social policy in harsh times: Social security development in Denmark, Finland, Norway and Sweden during the 1980s and 1990s," *International Journal of Social Welfare*, 9: 31–42.

Norlin, E. (2003) "An analysis of immigrants' participation in adult education in Sweden," *Umeå Economic Studies*, 617: 1–26.

Nygård, M. (2006) "Welfare-ideological change in Scandinavia: A comparative analysis of partisan welfare state positions in four Nordic countries, 1970–2003," *Scandinavian Political Studies*, 29(4): 356–385.

O'Farrell, C. (2005) *Michel Foucault*. London: Sage Publications.

OECD (2000) *International Migration Statistics*. Online. Available HTTP: www.source-oecd.org.

OECD (2002) *Babies and bosses: Reconciling work and family life*. Paris: OECD.

OECD (2004) *Trends in international migration: Sopemi 2003*. Paris: OECD.

OECD (2005) *Social expenditure database*. Online. Available HTTP: www.oecd.org/dataoecd/20/18/39696251.pdf.

OECD (2007a) *Trends in international migration*. Paris: OECD.

OECD (2007b) *Babies and bosses—Reconciling work and family life: A synthesis of findings for OECD countries*. Paris: OECD.

OECD (2008) *SF3. Fertility rates. Family database*. Online. Available HTTP: www.oecd.org/els/social/family/database.

OECD (2009a) *OECD factbook 2009: Economic, environmental and social statistics*. Paris: OECD.

OECD (2009b) *International migration outlook: SOPEMI 2009*. Paris: OECD.

OECD (2010) *Social expenditure database*. Online. Available HTTP: http://stats.oecd.org/Index.aspx?DatasetCode=CSP2010.

Ojakangas, M. (2005) "Impossible dialogue on bio-power: Agamben and Foucault," *Foucault Studies*, 2: 5–28.

Oláh, L.Sz. (2001) "Gender and family stability: Dissolution of the first parental union in Sweden and Hungary," *Demographic Research*, 4: 29–96.

Orloff, A. (1993) "Gender and the social rights of citizenship: The comparative analysis of gender relations and welfare states," *American Sociological Review*, 58: 303–328.

Østergaard, U. (1992) "Peasants and Danes: The Danish national identity and political culture," *Comparative Studies in Society and History*, 34(1): 3–27.

Østergaard-Nielsen, E. (2003) "The politics of migrants' transnational political practices," *International Migration Review*, 37: 760–786.

Palermo, G., Joris, H., Devroey, P., and Van Steirteghem, A.C. (1992) "Pregnancies after intracytoplasmic injection of single spermatozoon into an oocyte," *Lancet*, 340: 17–18.

Parry, J.T. (2005) " 'Society must be [regulated]': Biopolitics and the commerce clause in Gonzales v. Raich," *Lewis and Clark Law Review*, 9(4): 853–877.

Pastore, N. (1949) *The nature-nurture controversy*. New York: King's Crown Press.

Patel, R. and McMichael, P. (2004) "Third Worldism and the lineages of global fascism: The regrouping of the global South in the neoliberal era," *Third World Quarterly*, 25: 231–254.

Paul, D.B. (1984) "Eugenics and the Left," *Journal of the History of Ideas*, 45: 567–590.

Paul, D.B. (1995) *Controlling Human Heredity: 1865 to the Present*. Atlantic Highlands, NJ: Humanities Press.

Paul, D.B. (1998) *The politics of heredity*. Albany: State University of New York Press.

Paul, D.B. and Spencer, H.G. (1998) "The failure of a scientific critique: David Heron, Karl Pearson and Mendelian eugenics," *British Journal for the History of Science*, 31: 441–452.

Pearson, K. (1892) *The grammar of science*. London: Dent. Reprinted 1951.

Pearson, K. (1900) *The grammar of science*. London: Adam and Charles Black.

Pease, E. (1911) *The history of the Fabian Society*. New York: E.P. Dutton and Company Publishers.

Peled, Y. (1992) "Ethnic democracy and the legal construction of citizenship: Arab citizens of the Jewish State," *American Political Science Review*, 86: 432–443.

Pellegrino, E.D. (2000) "Bioethics at century's turn: Can normative ethics be retrieved?" *The Journal of Medicine and Philosophy*, 25: 655–675.

Peltoniemi, T. (2011) "Med ansvar for Europas totale misbruk," *Samfunn*, 2: 13–15.

Penslar, D.J. (1991) *Zionism and technocracy: The engineering of Jewish settlement in Palestine, 1870–1918*. Bloomington, IN: Indiana University Press.

Philo, C. (2001) "Accumulating populations: Bodies, institutions and space," *International Journal of Population Geography*, 7: 473–490.

Pick, D. (1989) *Faces of degeneration: A European disorder, c. 1848—c. 1918*. Cambridge: Cambridge University Press.

Pickens, D.K. (1968) *Eugenics and progressives*. Nashville: Vanderbilt University Press.

Pierson, P. (1994) *Dismantling the welfare state? Reagan, Thatcher, and the politics of retrenchment*. Cambridge: Cambridge University Press.

Pierson, P. (2001) "Post-industrial pressures on the mature welfare states," in P. Pierson (ed.) *The new politics of the welfare state* (pp. 80–104). New York: Oxford University Press.

Pitkänen, P. and Kouki, S. (2002) "Meeting foreign cultures: A survey of the attitudes of Finnish authorities towards immigrants and immigration," *Journal of Ethnic and Migration Studies*, 28(1): 103–118.

Plato (1974) *The republic* (2nd edn.), trans. D. Lee. Harmondsworth: Penguin.

Plows, A. (2011) *Debating human genetics*. London, New York: Routledge.

Population Reference Bureau (August 16, 2007) *World population data sheet*. Washington. DC: National Press Club. Online. Available HTTP: www.prb.org/Publications/Datasheets/2007/2007USPopulationDataSheet.aspx.

Porter, D. (1996) "Social medicine and the new society: Medicine and scientific humanism in mid-twentieth century Britain," *Journal of Historical Sociology*, 9: 168–187.

Portin, P. and Saura, A. (1985) *The impact of Mendel's work on the development of genetics in Finland*. Folia Mendeliana 20, Supplementum ad Acta Musei Moravie Scientiae Naturales LXX (Brno: Moravske Musei Moravie).

Portugese, J. (1998) *Fertility policy in Israel: The politics of religion, gender, and nation*. Westport, CT: Praeger.

Presner, T.S. (2003) "Clear heads, solid stomachs, and hard muscles: Max Nordau and the aesthetics of Jewish regeneration," *Modernism/Modernity*, 10(2): 269–296.

Prioux, F. (2003) "Age at first union in France: A two-stage process of change," *Population-E*, 58(4–5): 559–578.

Prioux, F. (2005) "Recent demographic developments in France," *Population-E*, 60(4): 371–414.

Proctor, R.M. (1988) *Racial hygiene: Medicine under Nazis*. Cambridge: Harvard.

Public Policy Institute of California (2002) *How fertility changes across immigrant generations. Research Brief* No. 58. San Francisco, CA: PPIC.

Rabinow, P. and Rose, N. (2006) "Biopower today," *BioSocieties*, 1: 195–217.

Ray, L.J. (1983) "Eugenics, mental deficiency and Fabian socialism between the wars," *Oxford Review of Education*, 9(3): 213–222.

Raz, A.E. (2009) "Eugenic utopias/dystopias, reprogenetics, and community genetics," *Sociology of Health & Illness*, 31(4): 602–616.

Richard, A.-M. and Mahon, R. (2001) "Sweden: Model in crisis," in J. Jenson and M.

Sineau (eds.) *Who cares? Women's work, childcare, and welfare state redesign*. Toronto: University of Toronto.

Richardson, G. (1994) *A welcome for every child: How France protects maternal and child health*. Arlington, VA: National Center for Education in Maternal and Child Health.

Roberts, M. (1938) *Bio-politics: An essay in the physiology, pathology and politics of the social and somatic organism*. London: Dent Ruse.

Roll-Hansen, N. (1996) "Conclusions: Scandinavian eugenics in the international context," in G. Broberg and N. Roll-Hansen (eds.) *Eugenics and the welfare state: Sterilization policy in Denmark, Sweden, Norway and Finland*. Michigan: Michigan State University Press.

Rothmayr, C., Varone, F., and Montpetit, E. (2003) "Does federalism matter for biopolicies? Switzerland in comparative perspective," *Swiss Political Science Review*, 9(1): 109–136.

Rothmayr, C., Varone, F., Serdült, U., Timmermans, A., and Bleiklie, I. (2004) "Comparing policy design across countries: What accounts for variation in ART policy?" in I. Bleiklie, M. Goggin, and C. Rothmayr (eds.) *Governing assisted reproductive technology: A cross-country comparison* (pp. 228–253). London: Routledge.

Rothstein, B. (1988) *Social classes and political institutions: The roots of Swedish corporatism*. Uppsala: Maktutredningen.

Rubin, I. (1933) "Ingathering of the exiles from eugenic perspective," *Moznaim*, 1(4): 89–93 [Hebrew].

Ruppin, A. (1904) *Die Juden der Gegenwart: Eine soxialwissdenschaftlicheStudie*. Berlin: Auflage (Unknown Binding).

Ruppin, A. ([1904]1913) *The Jews of Today*, trans. M. Bentwich. New York: H. Holt and Company.

Ruppin, A. (1934) *The sociology of the Jews* (2nd edn.), 2 vols. 4 parts. Tel-Aviv: A.Y. Shtibel [Hebrew].

Ruppin, A. (1936) "The selection of the fittest," in *Three decades of Palestine: Speeches and papers on the upbuilding of the Jewish national home*. Tel-Aviv: Greenwood Press.

Sainsbury, D. (1996) *Gender, equality and welfare states*. Cambridge: Cambridge University Press.

Sandier S., Paris, V., and Polton, D. (2004) *Health care systems in transition: France*. Brussels, Belgium: WHO Regional Office for Europe on behalf of European Observatory on Health Systems and Policies.

Sapp, J. (1987) *Beyond the gene: Cytoplasmic inheritance and the struggle for authority in genetics*. New York: Oxford University Press.

Sarig, M. (November 1, 2002) "We love you birth-mothers," *Ha'aretz*, pp. 15–16.

Satka, M. (1995) *Making social citizenship: Conceptual practices from the Finnish Poor Laws to professional social work*. Publication of Social and Political Sciences and Philosophy. University of Jyvaskyla.

Satka, M. (2009) "Investing in future citizens: Finnish social policies and child welfare social work after World War II," in G. Haus and D. Schulte (eds.) *Amid social contradictions, towards a history of social work in Europe*. Opladen and Farmington: Barbara Budrich Publisher.

Scharpf, F. (2000) "Economic changes, vulnerabilities and institutional capabilities," in F. Scharpf and V. Schmidt (eds.) *Welfare and work in the open economy*. Vol. I (pp. 21–124). Oxford: Oxford University Press.

Schiffino, N. and Varone, F. (2004) "Art policy in Belgium: A bioethical paradise?" in I. Bleiklie, M. Goggin, and C. Rothmayr (eds.) *Governing assisted reproductive technology: A cross-country comparison* (pp. 21–41). London: Routledge.

Schiller, F.C.S. (1921) "Eugenics versus civilization," *The Eugenics Review*, 13: 381–393.

Schiller, F.C.S. (1925) "The ruin of Rome and its lessons for us," *The Eugenics Review*, 17: 4–14.

Schmidt, V. (2000) "Values and discourse in the politics of adjustment," in F. Scharpf and V. Schmidt (eds.) *Welfare and work in the open economy*. Vol. I (pp. 229–319). Oxford: Oxford University Press.

Schmidt, V. (2002) "Does discourse matter in the politics of welfare state adjustment?" *Comparative Political Studies*, 35: 168–193.

Schnapper, D. (1994) *La Communaut´e des citoyens: Sur l'id´ee moderne de nation* [The community of citizens: On the modern idea of nation]. Paris: Gallimard.

Schulze, H.J. (2003) *Netherlands. Family observer*. Luxembourg: Office for the Official Publications of the European Communities.

Sciebinger, L. (1989) "The anatomy of differences: Race and sex in eighteenth century science," *The Eighteenth Century Studies*, 22: 387–405.

Searle, G.R. (1976) *Eugenics and politics in Britain 1900–1914*. Leyden: Noordhoff International Publishing.

Semmel, B. (1958) "Karl Pearson: Socialist and Darwinist," *British Journal of Sociology*, 9: 111–125.

Semmel, B. (1960) *Imperialism and social reform: English social-imperial thought 1895–1914*. London: Allen and Unwin.

Shadid, W.A.R. and van Koningsveld, P.S. (eds.) (1995) *Religious freedom and the neutrality of the state: The position of Islam in the European Union*. Leuven: Peeters.

Shafir, G. (1989) *Land, labor and the origins of the Israeli-Palestinian conflict, 1882–1914*. Cambridge: Cambridge University Press.

Shalev, C. and Gooldin, S. (2006) "The uses and misuses of IVF in Israel: Some sociological and ethical considerations," *Nashim*, 12: 151–176.

Shaw, G.B. (1891) "The transition to Social Democracy," in H.G. Wilshire (ed.) *Fabian Essays in Socialism*, American Edition (pp. 213–246). New York: The Homboldt Publishing Co.

Shaw, G.B. (1904) "On the history of Fabian economics", in E.R. Pease, *The History of the Fabian Society*. New York: E. Dutton and Company Publishers.

Shvarts, S., Davidovitch, N., Seidelman, R., and Goldberg, A. (2005) "Medical selection and the debate over mass immigration in the new state of Israel (1948–1951)," *Canadian Bulletin of Medical History*, 22(1): 5–34.

Simons, M. (2006) "Learning as investment: Notes on governmentality and biopolitics," *Educational Philosophy and Theory*, 38(4): 523–540.

Skovmand, K. (1996) "Danish fertility bill sparks debate," *The Lancet*, 347: 462.

Smart, B. (1985) *Michel Foucault*. London and New York: Routledge.

Smythe, I. (1994) *Population policies: Official responses to feminist critiques*. DP-14. London: School of Economics, Centre for the Study of Global Governance.

Soder, H.-P. (1991) "Disease and health as contexts of modernity: Max Nordau as a critic of fin-de-siècle modernism," *German Studies Review*, 14(3): 473–487.

Soloway, R.A. (1990) *Demography and degeneration*. Chapel Hill, NC: University of North Carolina Press.

Somit, A. (1972) "Biopolitics," *British Journal of Political Science*, 2(2): 209–238.

Sorel, G. (1976) "The socialist future of the syndicates," in *From Georges Sorel: Essays in socialism and philosophy*, trans. J. Stanley and C. Stanley (pp. 71–93). New York: Oxford University Press.

Sorel, G. (1999) *Reflections on violence*, ed. J. Jennings. Cambridge: Cambridge University Press.

Sorenson, C. (2006) "IVF/ART funding and reimbursement," *Euro Observer*, 8(3): 6–7.

Soysal, Y. (1996) *Limits of citizenship: Migrants and post national membership in Europe*. Chicago: University of Chicago Press.

Soysal, Y. (1997) "Changing parameters of citizenship and claims-making: Organized Islam in European public spheres," *Theory and Society*, 26(4): 509–527. Special Issue on Recasting Citizenship.

Sparke, M. (2006) "A neoliberal nexus: economy, security and the biopolitics of citizenship on the border," *Political Geography*, 25: 151–180.

Spektorowski, A. (2004) "The eugenic temptation in socialism," *Comparative Studies in History and Society*, 46(01): 84–106.

Spektorowski, A. and Mizrachi, E. (2004) "Eugenics and the welfare state in Sweden: The politics of social margins and the idea of a productive society," *Journal of Contemporary History*, 39: 333–352.

Spencer, H. (1966) "Social static, abridged and revised (1884)," in *Works of Herbert Spencer*, Vol. 21 (pp. 147–148). Osnabrock: Otto Zeller.

Spengler, O. (1919) *Prussiandom and Socialism*. In Donald O. White (ed.) *Oswald Spengler, Selected Essays*. Henry Regnery Co., 1967.

Stalker, P. (2002) "Migration trends and migration policy in Europe," *International Migration*, 40(5): 151–179.

Starobin, P.K. (2004) "Going Dutch," *National Journal* (February 28). *General Family Monitoring Report 2003: The Netherlands*. European Union Family Observatory.

Statens medicinsk-etiska råd (1995) Statens medicinsk-etiska råd: Assisterad befruktning—Synpunkter på vissa frå-gor i samband med befruktning utanför kroppen. Statens medicinsk-etiska råd 1995–04–05. Online. Available HTTP: www.smer.gov.se/index. htm?lang=sv&index=4&url=yttranden.html.

Statistics Denmark (2009) Online. Available HTTP: www.statbank.dk/statbank5a/default.asp?w=1024.

Statistics Finland (1997) *Population*. Online. Available HTTP: www.stat.fi/tup/astika/kaytto/astika30011_en.html.

Statistics Finland (2000) *Aliens and international migration 1999*. Helsinki: Statistics Finland Population.

Statistics Finland (April 2009) Online. Available HTTP: www.stat.fi/til/synt/2008/synt_2008_2009–04–22_tie_001_en.html.

Stenius, H. (1997) "The good life is a life of conformity: The impact of the Lutheran tradition on Nordic political culture," in O. Sorenson and B. Stråth (eds.) *The cultural construction of Norden*. Oslo: Scandinavian University Press.

Stepan, N. (1991) *The hour of eugenics: Race, gender and nation in Latin America*. Ithaca: Cornell University Press.

Sternhell, Z. (1979) "Fascist ideology," in W. Laqueur, *Fascism: A reader's guide: Analyses, interpretations, bibliography* (pp. 315–378). New York: Oxford University Press.

Sternhell, Z. (1998) *The founding myths of Israel: Nationalism, socialism, and the making of the Jewish State*. Princeton, NJ: Princeton University Press.

Sternhell, Z., Sznajder, M., and Asherri, M. (1994) *The birth of Fascist ideology*. Princeton, NJ: Princeton University Press.

Stocking G.W. (1968) *Race culture and evolution: Essays in the history of anthropology*. New York: Free Press.

Stoler-Liss, S. (2003) "Mothers birth the nation: The social construction of zionist motherhood in wartime in Israeli parents' manuals I," *Nashim: A Journal of Jewish Women's Studies & Gender Issues*, 6: 104–118.

Stoler-Liss, S. and Shvarts, S. (2004) "Fighting ignorance and undeveloped habits: Doctor's and nurses' conceptions regarding immigrants during mass immigration to Israel, 1949–1956," *Israel*, 6: 31–62 [Hebrew].

Stone Sweet, A. (2000) *Governing with judges: Constitutional politics in Europe*. Oxford: Oxford University Press.

Sufian, S. (2007) "Mental hygiene and disability in the Zionist project," *Disability Studies Quarterly*, 27(4); Online. Available HTTP: www.dsq-sds.org.

Sum, N.L. (2004) *From "integral state" to "integral world economic order": Towards a Neo-Gramscian cultural international political economy*. Cultural Political Economy Working Paper Series; Working Paper No. 7, University of Lancaster.

Sunde, A. (2007) "Europe's declining population and the contribution of ART," *Pharmaceuticals Policy and Law*, 9: 79–89.

Svirsky, S., Kraus, V., Konor-Attias, E., and Herbst, A. (2002) *Single mothers in Israel*. Tel-Aviv: Advah Centre.

Swank, D. (2006) "Tax policy in an era of internationalization: explaining the spread of neoliberalism," *International Organization*, 60(4): 847–882.

Swedish Institute (April 2003) *Factsheets on Sweden—the Swedish population*. Online. Available HTTP: www.si.se/.

Swedish Integration Board (2007) Online. Available HTTP: www.integrationindex.eu/integrationindex/2549.html.

Swedish Migration Board Statistik (2001) *Statistics Sweden*. Online. Available HTTP: http://scb.se/sm/BO34SM0101_tabeller12.asp.

Taskinen, S. (2003) *The situation of families in Finland*. European Observatory on the Social Situation, Demography, and Family. Online. Available HTTP: http://europa.eu.int/comm/employment_social/eoss/index_en.html.

Tawat, M. (2011) "A comparative study of the immigrant integration policies of Denmark and Sweden, 1960–2006," (Ph.D. diss., University of Otago, Dunedin, New Zealand).

Taylor-Gooby, P. (2001) "Risk, contingency and the Third Way: Evidence from the BHPS and qualitative studies," *Social Policy and Administration*, 35(2): 195–211.

Taylor-Gooby, P. (2004) "Open markets and welfare values," *European Societies*, 6: 29–48.

Timmermans, A. (2003) *High politics in the low countries*. Aldershot: Ashgate.

Timmermans, A. (2004) "The Netherlands: Conflict and consensus on assisted reproductive technology policy," in I. Bleiklie, M. Goggin, and C. Rothmayr (eds.) *Comparative biomedical policy: Governing assisted reproductive technologies* (pp. 155–173). London: Routledge.

Timonen, V. (2003) *Restructuring the welfare state*. Cheltenham: Edward Elgar.

Titmuss, R.M. (1974) *Social policy: An introduction*. London: Allen and Unwin.

Torfing, J. (2001) "Path-dependent Danish welfare reforms: The contribution of the new institutionalisms to understanding evolutionary change," *Scandinavian Political Studies*, 24: 277–309.

Törnudd, K. (1996) "Ties that bind to the recent past—Debating security policy in Finland within the context of membership of the European Union," *Cooperation and Conflict, Nordic Journal of International Studies*, 31(1): 37–68.

Tranaes, T. and Zimmermann, K.F. (2004) *Migrants, work, and the welfare state*. Odense: University Press of Southern Denmark and the Rockwool Foundation Research Unit.

Trattner, W.I. (1979) *From poor law to welfare state: A history of social welfare in America*. New York: Free Press.

Tremain, S. (2005) "Foucault, governmentality, and critical disability theory: An introduction," in *Foucault and the government of disability*. Ann Arbor: University of Michigan Press.

Tremain, S. (2006) "Reproductive freedom, self-regulation, and the government of impairment in utero," *Hypatia: A Journal of Feminist Philosophy*, 21(1): 35–43.

Tremain, S. (2008) "The biopolitics of bioethics and disability," *Bioethical Inquiry*, 5: 101–116.

Trombley, S. (1988) *The right to reproduce: A history of coercive sterilization*. London: Weidenfeld and Nicolson.

Turda, M. (2005) *The idea of national superiority in Central Europe, 1880–1918*. New York: Oxford University Press.

Turda, M. and Weindling, P.J. (eds.) (2007) *Blood and homeland: Eugenics and racial nationalism in Central and Southeast Europe, 1900–1940*. Budapest and New York: Central European University Press.

Tydén, M. (2002) *Från politik till praktik. De svenske sterilizeringslagarna 1935–1975* [From policy to practice. The Swedish sterilization acts 1935–1975], (2nd expanded edition) Stockholm Studies in History—63, Acta Universitatis Stockholmiensis, Stockholm University.

UNICEF (2000) *Innocenti report card, 1*. Florence, Italy: UNICEF.

UNICEF (2005) *Child poverty in rich countries*. Online. Available HTTP: www.euro.centre.org/data/1182414404_23488.pdf.

UN/ECE (United Nations Economic Commission for Europe) (1999) *Population in Europe and North America on the eve of the millennium: Dynamics and policy responses*. National report submitted by the Government of Sweden. Geneva and New York: United Nations Economic Commission for Europe.

Valtonen, K. (2001) "Cracking monopoly: Immigrants and employment in Finland," *Journal of Ethnic and Migration Studies*, 27(3): 421–438.

Van Eeckhout, L. (April 24, 2006) "Les Eglises chrétiennes s'inquiètent du contenu du projet de loi sur l'immigration," *Le Monde*.

Van Gunsteren, H. (March 19, 2010) "Stop het Normaliseren van de Burger" [Stop the normalization of the civilian], *De Volkskrant*.

Varone, F., Rothmayr, C., and Montpetit, E. (2006) "Regulating biomedicine in Europe and North America: A qualitative comparative analysis," *European Journal of Political Research*, 45(27): 317–343.

Virtanen, A.-M. (1985) "Gobineau's racial doctrine and the idealization of Germanic peoples," in A. Kemiläinen (ed.) *Mongols or Germanics?* Helsinki: Finnish Historical Society.

Volden, C. (2006) "States as policy laboratories: Emulating success in the children's health insurance program," *American Journal of Political Science*, 50(2): 294–312.

Webb, S. (1909) "Eugenic and the poor laws," *The British Medical Journal*, December 25, p. 1808.

Webb, S. (1912) "The economic theory of a legal minimum wage," *Journal of Political Economy*, 20(10): 973–998.

Webb, S. and Webb, B. ([1894]1976) *History of trade unionism*. New York: AMS Press.

Webb, S. and Webb, B. ([1897]1920) *Industrial democracy*. London: Longmans, Green, and Co.

Weingart, P. (1999) "Science and political culture: Eugenics in comparative perspective," *Scandinavian Journal of History*, 24: 163–177.

Weiss, D. (2000) "The European Union and the family," in A. Pfenning and T. Bahle (eds.) *Families and family policies in Europe: Comparative perspectives* (pp. 127–140). Frankfort: Peter Lang.

Weiss, S.F. (1990) "The race hygiene movement in Germany, 1904–1945," in M. Adams (ed.) *The wellborn science*. Oxford: Oxford University Press.

Wells, H.G. (1902) *Anticipations of the reaction of mechanical and scientific progress upon human life and thought*. London, New York: Harper and Brothers.

Wells, H.G. (1906) *Socialism and the family*. London: A.C. Fifield.

Wells H.G. (1908) *A modern utopia*. London: Odhams Press Limited.

Wells, H.G. (2004) "Introduction," in Margaret Sanger, *The pivot of civilization*. Kila, MT: Kessinger Publishing.

Wertz, D.C. (1998) "Eugenics is alive and well: A survey of genetic professionals around the world," *Science in Context*, 11: 493–510.

Wihtol de Wenden, C. (2003) "Assimilation and struggle," *Georgetown Journal of International Affairs*, 4(2): 69–75.

Wilkinson, S. (2006) "Eugenics, embryo selection and the equal value principle," *Clinical Ethics*, 1: 46–51.

Williams, F. (2001) "Race/ethnicity, gender and class in welfare states: A framework for comparative analysis," in J. Fink, G. Lewis, and J. Clarke (eds.) *Rethinking European welfare* (pp. 131–163). Thousand Oaks, CA: The Open University Press.

Willis, E. (2005) "Public health and the 'new genetics'," in R. Bunton and A. Petersen (eds.) *Genetic governance: Health, risk and ethics in the biotech era* (pp. 155–170). Oxford: Routledge.

Wright, S. (1960) "Genetics and twentieth century Darwinism," *American Journal of Human Genetics*, 12: 368.

Wright, S. (1968) *Evolution and the genetic of populations*, Vol. I. Chicago and London: University of Chicago Press.

Yassky, C. (1937) "Hachazarat hacholim hashuchei marpe lechutz leAretz," *Bulletin on Social Welfare in Palestine*, 3–4: 54–55.

Yuval-Davis, N. (1997) *Gender and nation*. London: Sage.

Zalashik, R. and Davidovitch, N. (2009) "Professional identity across the borders: Refugee psychiatrists in Palestine, 1933–1945," *Social History of Medicine*, 22(3): 569–587.

Zaremba, M. (August 21, 1997) "De olönsamma skars bort" [The unprofitable were chopped off], *Dagens Nyheter*.

Zaremba, M. (August 20, 1997) "Rasren i välferdsstaten" [No mixing of races in the Welfare State], *Dagens Nyheter*.

Zweiniger-Bargielowska, I. (2007) "Raising a nation of 'good animals': The new health society and health education campaigns in interwar Britain," *Social History of Science*, 20: 73–89.

Index

Page numbers in *italics* denote tables, those in **bold** denote figures.